Economic Circularity in the Roman and Early Medieval Worlds

New Perspectives on Invisible Agents and Dynamics

Proceedings of the conference held at the Centre for Urban Network Evolutions/Moesgaard Museum, Aarhus (7–8 September 2022)

Edited by

Irene Bavuso, Guido Furlan,
Emanuele E. Intagliata and Julia Steding

OXBOW | books
Oxford & Philadelphia

Published in the United Kingdom in 2024 by
OXBOW BOOKS
The Old Music Hall, 106-108 Cowley Road, Oxford, OX4 1JE

and in the United States by
OXBOW BOOKS
1950 Lawrence Road, Havertown, PA 19083

Hardback edition: ISBN 978-1-78925-996-4
Digital Edition: ISBN 978-1-78925-997-1 (epub)

A CIP record for this book is available from the British Library

Library of Congress Control Number: 2023944919

Printed in the United Kingdom by Short Run Press
Typeset in India by DiTech Publishing Services

For a complete list of Oxbow titles, please contact:

UNITED KINGDOM
Oxbow Books
Telephone (0)1226 734350
Email: oxbow@oxbowbooks.com
www.oxbowbooks.com

UNITED STATES OF AMERICA
Oxbow Books
Telephone (610) 853-9131, Fax (610) 853-9146
Email: queries@casemateacademic.com
www.casemateacademic.com/oxbow

Oxbow Books is part of the Casemate Group

Front cover: Top left: Palmyra, Peristyle Building (Pal.M.A.I.S./University of Milan).
Top right: Fountain of the house of the Grand Duke of Tuscany, detail of the mosaic decoration with sone, Egyptian blue, glass tesserae, sea shells and fragments of twisted glass rods, second quarter first century CE, Pompeii (photo Cristina Boschetti, courtesy Soprintendenza Archeologica di Pompei). This image is published: Cristina Boschetti, "Vitreous Materials in Republican and Early Imperial Mosaics (2nd century BCE–1st century CE) from Italy: Faience, Egyptian Blue, and Glass." Journal of Glass Studies 53 (2011), 59–91, fig. 28.
Bottom left: Detail of wall mosaic combining shells, lumps of Egyptian blue and a fragment of blue glass. Segni (Roma), nymphaeum, late 2nd century BC (image from Boschetti C., 2020. Vetro e blu egizio nel ninfeo di Segni: aspetti decorativi, tipologici e tecnologici, in F.M. Cifarelli (ed.), Il ninfeo di Segni. Studi e ricerche, Roma, pp. 55–67).
Bottom right: Textile with cord imprint that was reused in waterproofing, found in situ on the hull of the Comacchio shipwreck. Photo by Margarita Gleba, courtesy Museo Delta Antico, Comacchio.

Contents

List of contributors

Editors (in alphabetical order by surname)

Irene Bavuso is Assistant Professor at the Department of History and Art History of Utrecht University. She is an early medieval historian, and her research concentrates on networks of trade, production, mobility and circulation of technical knowledge. She addresses these themes through a comparative and interdisciplinary approach across post-Roman northwestern Europe. She took a D.Phil. at the University of Oxford and held post-doctoral positions at the Institute of Historical Research (London), the Université libre de Bruxelles, the DFG-funded 'Migration and Mobility in Late Antiquity and the Early Middle Ages' Centre for Advanced Studies at the Eberhard Karls Universität Tübingen, and the Centre for Urban Network Evolutions (UrbNet), Aarhus University.

Guido Furlan is Assistant Professor in Archaeological Research Methodologies at the University of Padova. His research interests include the archaeology of Roman towns, and theoretical and methodological approaches to the study of the archaeological record.

Emanuele E. Intagliata is an Assistant Professor at the University of Milan, where he teaches Christian and medieval archaeology. His main research interests lie in the transformation of the townscapes and the life of frontier communities in Syria, Anatolia and western Georgia from the 4th to the 6th centuries AD. He has approached these topics through archaeological fieldwork and the study of legacy data. He is the author of the monograph *Palmyra after Zenobia* (Oxbow 2018) and numerous articles dedicated to the eastern Roman frontier in Late Antiquity and beyond.

Julia Steding is a Research Assistant at the Centre for Urban Network Evolutions, Aarhus University. She obtained her doctoral degree in 2020 at Aarhus University, and her thesis focused on the stone carving techniques and the organization of the portrait production in Roman Palmyra. Afterwards she held a post-doctoral position at the Centre for Urban Network Evolutions, Aarhus University. Her field of research is Roman funerary art from Italy and the Near East. She furthermore explores economic processes linked to funerary portraiture and architecture and the reuse of materials in these contexts.

Contributing authors (in alphabetical order by surname)

Cristina Boschetti (Ph.D. 2009, University of Padua) is a postdoctoral research fellow at UrbNet, Aarhus University, Denmark. She has previously held post-doctoral positions at the Universities of Nottingham (UK), Padua (IT) and Cairo (Egypt) and at CNRS IRAMAT-CEB, Orléans (FR). She has curated exhibitions on ancient jewellery. Her research interests include glass typology and the application of archaeometric analyses to investigate the economy and technology of glass, jewellery, mosaics, and paintings, with a focus on the period from Hellenism to the Middle Ages and on the development of long-term phenomena. ORCID: 0000-0003-4499-779X.

Maria Stella Busana is Associate Professor of Roman Archaeology at the Department of Cultural Heritage, University of Padua, Italy. Her research focuses on Roman archaeology and Roman economics. She has directed numerous archaeological excavations and surveys in Italy (Pompei, Altinum-Ca'Tron, Montebelluna), and many research projects on Roman textile economy (Pondera, TRAMA, Lanifica).

Rowan S. English is a Ph.D. fellow at the Centre for Urban Network Evolutions at Aarhus University. His main research interests are in the global archaeology of the 1st millennium CE. His current project focuses on the movement of elephant ivory from its source to graves around the North Sea in the 5th–8th centuries CE. Other research interests include ivory from all periods, the funerary archaeology of early medieval Europe, early medieval osseous objects and archaeological theory.

Jon Frey is a classical archaeologist specializing in Greek and Roman architecture, with a particular interest in construction processes and the phenomenon of material reuse in post-classical antiquity. While the majority of his fieldwork experience is in the area of Greece known as the Corinthia, over the past two decades he has also worked in Egypt and on the islands of Crete and Kythera. As the Director of the Michigan State University Excavations at Isthmia, he led a team of researchers engaged in the use of digital tools to explore and publish evidence about the ancient world contained in archaeological archives. He is the author of *Spolia in Fortifications and the Role of the Common Builder in Late Antiquity* (2015) and the recipient of an NEH Digital Humanities Implementation Grant.

Margarita Gleba is Assistant Professor at the Department of Cultural Heritage, University of Padua, Italy. Before moving to Italy, she was research project manager at the Danish National Research Foundation's Centre for Textile Research at the University of Copenhagen (2005–2009), Marie Curie Fellow at the Institute of Archaeology University College London (2009–2011), principal investigator of the ERC project PROCON (2013–2019) and a lecturer at the Ludwig Maximilian University of Munich (2020–2021). Her research interests include the archaeology of the pre- and protohistoric Mediterranean and western Asia. Her special area of research is textile archaeology, including scientific analytical methods of fibre and textile investigation.

Rhiannon Garth Jones is a Ph.D. Fellow at the Centre for Urban Network Evolutions, Aarhus University and an experienced communicator of complex ideas across a range of disciplines and through a variety of media, as most appropriate for the audience: academic papers, member journals, articles, blogs, podcasts, social media and public presentations. Her interests range from early Abbasid history; urban transformation in the Mediterranean and Near East; urban networks from the Hellenistic period to 10th century; the long late antiquity; artistic, cultural, ideological, and religious interactions and transformations in the ancient world; and digital archival practice.

Allyson McDavid is Assistant Professor of Ancient Material and Visual Culture at Parsons School of Design at The New School in New York City, where she teaches architectural design history, theory and practice. She received her Ph.D. in 2015 in architectural history and archaeology from New York University, has a B.A. in architecture and a M.Arch. from the University of California at Berkeley. As an architect on sites of antiquity across the Mediterranean and Black Sea regions, her research focuses on the conservation, renovation and sustainability of historic built environments and the relevance of these histories to contemporary design.

Derek Parrott received his B.A. in anthropology from Northern Kentucky University in the United States in 2016 and completed his M.A. at the University of York in 2018. During his M.A., Derek specialized in the material culture of the Viking Age, with a particular focus on jewelry and metalwork. After receiving his degree, he returned to the US to work as a commercial archaeologist, as both a field technician and principal investigator for several different archaeological and engineering firms across the country. These projects included surveys for land developers, site excavations and cemetery relocations. In February of 2022, he moved to Denmark to begin a Ph.D. with the Centre for Urban Network Evolutions at Aarhus University. This project seeks to contextualize the corpus of nonferrous metal-casting mould fragments excavated during the 2016–2018 Northern Emporium project, using the novel method of 3D scanning in order to more effectively visualize the often difficult-to-discern object impressions they contain.

Line Van Wersch has a Ph.D. in medieval archaeology and archaeometry. She specialises in glass and ceramics. Along postdoctoral contracts she studied collections form Belgium, France, Germany, and the Netherlands. She is presently Associate Professor at Liege University and Marie-Curie fellow at the CNRS (Arscan UMR 7041) where she is working on pyrotechnologies in the early Middle Ages.

Alexis Wilkin is Professor of Medieval History at the University of Brussels (ULB), Director of the Research Unit SOCIAMM (Sociétés anciennes, Médiévales et modernes) and a member of the Royal Historical Commission of Belgium. He is a specialist of medieval economic history, working on the institutional framing of trade, food distribution, and town and countryside relations. He has also a peculiar interest in the use of archaeological proxies in the reconstruction of medieval exchanges networks.

Jonathan Wood is Research Associate (Archaeomaterials) in the Department of Archaeology, Classics and Egyptology at the University of Liverpool and a teaching fellow at University College London. His recent publications have focussed on provenancing silver from ore to object, investigating glass and glaze production and finding ways to identify ancient recycling in the archaeological record.

Introduction: approaching invisible circular economies through archaeological and historical sources

Irene Bavuso, Guido Furlan, Emanuele E. Intagliata and Julia Steding

This volume stems from the conference 'Invisible Circularity from the Roman Period to the Middle Ages', held at Moesgaard Museum (Aarhus, Denmark) in September 2022. The present proceedings, like the conference, aim to investigate unusual and little studied (invisible) aspects of past circular economic practices, investigating both the archaeological record and the textual sources in order to build a more holistic picture of ancient economies. The focus of this volume falls, therefore, on circular economic processes (forms of reuse, repair, curation, maintenance) and the agents associated with them. Both are often difficult to identify in archaeological and historical sources, and very little research on them has been conducted to date. Also investigated in this volume is the incompleteness of relevant datasets, which often makes these processes even more difficult to detect (when possible at all) for archaeologists and historians. Further discussion concentrates on the ways in which the scholarly community can approach data lacunae to reach a more integrated understanding of conservative and regenerative economic processes in the pre-modern period.

The volume adopts a case-study approach which focuses on Europe and the Near East and which gives prominence to specific classes of material (*e.g.*, textiles), processes (*e.g.*, repurposing), agents (*e.g.*, scavengers) and methodologies (*e.g.*, archive archaeology) that are generally overlooked in modern scholarship. To avoid the sectorialisation that has often characterised this topic and to explore economic circularity in a multidisciplinary way, the case studies were chosen to include works by both historians and archaeologists, the latter with expertise in 'hard science' and material culture. The chronological scope of the volume is intentionally wide and spans the Roman period through the early Middle Ages in an attempt to investigate the topic from a *longue durée* perspective and, therefore, to avoid the simplistic categorisations and labels that have characterised the topic until recently. In so doing, the selected essays provide a vivid insight into the complexity of approaching economic circularity in these areas and periods, as well as its untapped research potential.

All the contributions set out to add to the ongoing discussion of economic circularity that has emerged over recent years. Instead of assuming that products were manufactured, consumed and discarded (the so-called linear economic model), a growing body of research has shown that reuse practices were more common in the past than it had been assumed (*e.g.*, Peña 2007; Duckworth and Wilson 2020; Andrade and Raja 2023; Bavuso *et al.* 2023). Among specific materials, glass has taken centre stage in discussions regarding recycling (*e.g.*, Freestone 2015; Boschetti *et al.* 2016; Paynter and Jackson 2016; Sainsbury 2019; Degryse 2020; Duckworth 2020; Barfod *et al.* 2022), closely followed by metal (*e.g.*, Wilson 2007; Pollard *et al.* 2015; Bray 2020; Ponting 2020). Naturally, archaeometry is a major player here, and indeed the journal *Archaeometry* released a special issue in 2022 on the topic of recycling (Sainsbury and Liu 2022). Research on the reuse of stone is even more popular and dates further back; archaeologists and historians keep exploring the movement of building materials in and between sites (for examples see the recent volumes on architecture by Y Ng and Swetnam-Burland 2018 and on sculptures by Gaggadis-Robin and de Larquier 2019). This volume joins current scholarship in advocating for a comprehensive picture of circular practices in past societies (Y Ng and Swetnam-Burland 2018; Duckworth and Wilson 2020) and does so by assuming interdisciplinary, *longue durée* perspectives that include generally overlooked fields. Indeed, much remains to be done, and research questions and methodologies – in both history and archaeology – still need improvement (Bavuso *et al.* 2023).

The contributors to the present volume seek to answer questions regarding the economic impact of circular systems and their organisation, as well as perceptions and practices beyond the economy. The contribution and role of archaeoscience is debated, and discussions include how to utilise newer technologies like 3D-scanning and the benefits of reviewing previously published material culture to reassess its significance within a circular system. Invisible agents – those that do not leave traces and those that are understudied – are revealed through the investigation of archaeological and historical materials, adding knowledge about ancient societies and how they functioned as well as shifting the focus from the impact of the elite on the economy to the role of the people that actually kept the economy running.

Every contribution raises several points of interest, contributing to the challenges and potentials raised above. For the sake of simplicity, the papers follow a thematic line: the earliest chapters mostly focus on the people involved in reuse practices and the organisation of these practices. The focus then shifts to different classes of materials involved in circular activities, and different methodological approaches are discussed (historical, archaeological, archaeometric). The last part of the book is dedicated to buildings and building materials.

In chapter 1, Guido Furlan follows the activity of ragpickers and junk dealers from the contemporary world to the Roman age. These individuals are well known in modern and medieval Europe, where proper guilds regulated their business. The written sources describe a multifaceted world that included important traders as well as poor collectors of rags and scraps. The overall impact on the economy of second-hand, reused materials, however, was not negligible. A similar picture emerges when taking into consideration the Roman world. Glass, metalware and textiles, for instance, were largely reused. The available epigraphic, juridical, and literary sources cast some light on the people involved in these activities, while archaeological data point to their substantial economic impact: the use of resources was optimised, and waste was avoided as much as possible. Overall, the activity of junk dealers and ragpickers characterised with considerable continuity the social and economic life of urban and rural communities through the last two millennia.

In chapter 2, Irene Bavuso investigates the organisation of scavenging in cemeteries during the early medieval period. Research has now established that scavenging of *villae* and other Roman buildings could take the form of an organised activity. Burial grounds, however, were perceived differently: grave-robbing was harshly punished in early medieval and earlier laws, and written sources are clear that it was considered an act of desecration and pollution. Yet the reopening of graves is amply attested. Bavuso explores diverse cases from Ostrogothic Italy to post-Roman England to reflect on the possible coexistence of various levels of organisation, from top-down management involving political power to bottom-up activities, and on the moral justifications that needed to be deployed and negotiated to violate tombs.

In chapter 3, Rhiannon Garth Jones investigates economic circularity in Abbasid Samarra, with a specific focus on top-down mechanisms that operated at the level of the caliphs. The contribution encompasses a broad view of the subject, which is explored through a holistic approach considering different practices, materials and objects – ranging from monumental iron gates to fire bricks and luxurious textiles. Besides recycling and reuse, the inclusion of the category of 'relocation' is noteworthy as it adds to a complex assessment of the range and significance of medieval circularity. Moreover, this approach makes it clear that we should not understand motivations behind early medieval circularity as compartmentalised: even at the highest levels of society, symbolic meanings and intent could and did coexist with practical thinking and resource-management strategies.

In chapter 4, Line Van Wersch and Alexis Wilkin examine the recycling and reuse of glass through the combined study of archaeometry and history. The chapter first introduces glass-making techniques and their development from the Roman to the medieval periods, and then discusses the three fundamental uses of glass during the early Middle Ages – that is, the production of beads, vessels and architectural glass (*e.g.*, window glass and tesserae). Here, the case of Germigny-de-Prés enables the authors to appreciate the full potential of archaeometry for investigating reuse and recycling. Thanks to the technical study of the material and the careful examination of the written sources, Van Wersch and Wilkin delve into the complexity of early medieval circuits of production and distribution on multiple, entangled levels: networks of elites and monasteries, urban and rural centres, craftsmen and patrons – up to the very elusive practices of daily life.

The reuse of Roman and early medieval glass is usually associated with remelting by secondary workshops. In chapter 5, Cristina Boschetti investigates forms of reuse that have received less attention from the scholarly community. In an original and effective way, the discussion is arranged into three parts based on three different reasons for reuse – namely beauty, utility and value. In the first case, coloured glass sherds used in combination with seashells and tufa for wall mosaics to recall a natural, rocky environment are discussed. Reworked glass fragments well exemplify a utilitarian approach: they were employed as knife sharpeners or blades, and some of these 'odd objects', recovered from early medieval burials, may have been used to light fires. Finally, the third aspect, value, emerges through a consideration of the curation and long-distance trade of beads and spindle whorls between their first use in Late Antiquity and the early Middle Ages, and their probable secondary use as currency and ornaments, respectively.

Dealing with incomplete data is the core of archaeological enquiry. Chapter 6, by Jonathan Wood, provides a thought-provoking discussion of the fundamental role of glass and metal recycling in provenance analyses. Reversing a well-rooted perspective, the author points out that considering recycling should be a preliminary step in archaeometric analyses and not a simple explanation for blurred results. Discussing several case studies ranging from New Kingdom Egypt to Roman and Parthian Mesopotamia, Wood also advocates for the use of a multi-analytical approach, warning against relying solely on a single source of data. Only the reliable identification of recycled materials in the archaeological record can lead to a better knowledge of productive and trade chains, and therefore to the creation of a sound and less simplistic economic history.

In chapter 7, Margarita Gleba and Maria Stella Busana investigate the reuse of textiles in Roman naval contexts. Drawing from materials recovered in several sites, including Comacchio, Lyon, Arles, Nydam Mose and Nemi, the authors discuss the reuse of rags in shipbuilding: rags were extensively employed to waterproof hulls, for insulation and as tarring tools. The variability observed in weave types confirms that the rags were obtained from a variety of sources, including garments and sails. While at first this practice seems to affect a small economic niche, through a quantification of the fabrics needed per ship, Gleba and Busana demonstrate the importance of this circular economic practice, which likely involved specialised traders, such as the *centonarii*, whose *collegia* are well known thanks to the available epigraphic evidence.

Chapter 8, by Rowan English, discusses the functional, aesthetic and spiritual modalities of reuse. His case study looks at necklace pendants from 7th-century lowland Britain that were found in graves associated with elite women. Instead of seeing reuse as merely an economically driven practice, he demonstrates the functional and aesthetic aspects of the practice. The functional aspects are closely linked to economic reasons but are not identical. Aesthetic motivations for reuse can also be proven and were often related to the inability to create a similar object. Reusing an object or parts of an object (even if modified) allowed for the creation of something that was equally as aesthetically pleasing as the original object. The spiritual aspect can be seen through a carefully conducted selection process that reflects on the power of nature, the power of the elite or an object's history.

In chapter 9, Derek Parrott addresses metal-casting models for star-shaped brooches from Viking Age Scandinavia. After describing the process of casting and establishing a chronology, Parrott demonstrates the potential of 3D scanning for the reconstruction of moulds and how this can help to match moulds – only found as fragments – with brooches. This can then lead to a better chronology and an understanding of the reuse of models, the change of models and the reasons behind these changes.

Chapter 10 by Emanuele E. Intagliata explores archival collections and their potential to fill in the data lacuna on circular economic processes left by past excavations. He does so by using two case studies, namely, the documents associated with the excavations of the Sanctuary of Baalshamin in the Collart Archive of the University of Lausanne and those of the Dörner Archive at the Forschungsstelle Asia Minor - University of Munster. In the former, Intagliata explores documents associated with the reuse of architectural and sculptural elements for the construction of new buildings within the temenos of the sanctuary, while in the latter, he investigates potential recycling activities at Eski Kahta - Commagene - in the medieval period. Both case studies demonstrate the potential and difficulties of conducting research on circular economic processes using archival data and, therefore, the complexity of reaching a holistic perspective on this topic.

In chapter 11 by Allyson McDavid, the central focus is on the thermal architecture of the Roman Empire. Bathhouses are monumental buildings with long lifespans that provide proof of an institutional bathing tradition. At the same time, they also provide an excellent subject for the study of reuse and renovation at the intersection of architectural practice and the agents involved in it. McDavid provides an overview and examples of the architectural alterations to bathhouses and discusses the 'why' of circular practices. Moreover, she looks at how administrative structures, resource management and rituals influenced building activities and decisions about the repair, renovation and redefinition of thermal buildings throughout Late Antiquity.

Chapter 12 by Jon Frey explores the reuse of columns in the ancient world. Columns were difficult to reuse due to their size and shape. Yet, their re-employment was ubiquitous in Roman and Byzantine architecture. Consequently, they have become the epitome of *spolia* in modern scholarship. Written sources and archaeological evidence offer numerous examples of columns reused for their original purpose. However, when not reused as vertical supports, columns could find a 'linear end' to their lives, embedded into later walls. This linear perspective on reuse is explored by Frey through three case studies from Greece (Athens, Sparta and Orchomenos).

In chapter 13, Julia Steding presents a dataset of reused inscriptions from Palmyra. She uses the data to point out the gap in the material evidence and, at the same time, shows what the study of one particular group of material can contribute to the understanding of circular practices in a specific site. The spatial movement of stone objects is discussed, and the period of original use is compared to the period in which the objects was reused. While acknowledging the limitations of the dataset, Steding shows that many objects were moved within the same context and that reuse can be clearly seen in relation to the defensive structures that were erected in and after the late 3rd century AD.

What emerges from the contributions in this volume is that the invisibility of circular economic processes is very much the result of biased and targeted research that has often neglected the topic. When addressed on a larger scale, the past circular economy reveals itself to be diverse and multifaceted. It brings to light a wide range of well-established activities, involving a large variety of materials reused in unexpected ways – reworked glass, old rags for shipbuilding, inscriptions and columns as building material, high status artwork and everyday objects. Both well-known and unexpected agents played different roles at various stages of circular economic processes: humble junk dealers, tomb scavengers, higher status textile traders and even political or religious authorities. Finally, we can also see a range of different reasons for these activities, including economic convenience, aesthetics and deep cultural or political motivations.

All these aspects suggest that circular economic practices should not be considered as exceptional approaches in periods of economic crisis, but as deep-rooted activities that characterised past economic systems, albeit in different ways and to different degrees. The 'apparent invisibility' of these practices does not document their absence; on the contrary, it tells the story of their effectiveness through the ages. The great variability in processes that has emerged from these contributions also suggests the necessity of pursuing a more holistic approach to the ancient economy in the future: to include circular practices in the narrative of past economies, a more integrated and cross-disciplinary approach to the subject must be developed. With this volume we hope to push forward studies of circular economies in pre-modern societies and contribute to the study of sites and materials through the lens of reuse habits, demonstrating how these were an integral part of urban economies in the Roman and early medieval periods.

Acknowledgements

The conference and the publication of its proceedings were supported by the Carlsberg Foundation under the grant CF22-0013 and the Danish National Research Foundation under the grant DNRF119 – Centre of Excellence for Urban Network Evolutions. The editors would like to extend their gratitude to these institutions for making this event and publication possible and to Moesgaard Museum for having hosted the event.

We are also grateful to the staff at Oxbow for their interest in this editorial project and to Courtney A. Ward for proofreading the contributions in this volume.

Bibliography

Andrade, N. and Raja, R. (eds) (2023) *Exchange and Reuse in Roman Palmyra: Examining Economy and Circularity,* Studies in Palmyrene Archaeology and History 8. Turnhout, Brepols.

Barfod, G.H., Freestone, I.C., Jackson-Tal, R.E., Lichtenberger, A. and Raja, R. (2022) Exotic glass types and the intensity of recycling in the northwest Quarter of Gerasa (Jerash, Jordan), *Journal of Archaeological Science* 140. DOI: https://doi.org/10.1016/j.jas.2022.105546.

Bavuso, I., Intagliata, E.E., Furlan, G. and Steding, J. (2023) Circular Economy in the Roman Period and the Early Middle Ages. Methods of Analysis for a Future Agenda. *Open Archaeology* 9.1. doi: 10.1515/opar-2022-0301.

Boschetti, C., Gratuze, B., Cavalieri, M., Lenzi, S. and Schibille, N. (2022) Production or Consumption? Glass Beads from the Roman Villa of Aiano, Tuscany. *Journal of European Archaeology* 25.2, 196–215.

Boschetti, C., Mantovani, V. and Leonelli, C. (2016) Glass Coloring and Recycling in Late Antiquity: A New Case Study from Aquileia (Italy). *Journal of Glass Studies* 58, 69–86.

Bray, P. (2020) Modelling Roman Concepts of Copper-Alloy Recycling and Mutability: The Chemical Characterization Hypothesis and Roman Britain. In C.N. Duckworth and A. Wilson (eds) *Recycling and Reuse in the Roman Economy*, 237–264. Oxford, Oxford University Press.

Degryse, P. (2020) Elements, Isotopes, and Glass Recycling. In C.N. Duckworth and A. Wilson (eds) *Recycling and Reuse in the Roman Economy*, 285–300. Oxford, Oxford University Press.

Duckworth, C.N. (2020) Seeking the Invisible: New Approaches to Roman Glass Recycling. In C.N. Duckworth and A. Wilson (eds) *Recycling and Reuse in the Roman Economy*, 301–356. Oxford, Oxford University Press.

Duckworth, C.N. and Wilson, A. (eds) (2020) *Recycling and Reuse in the Roman Economy*. Oxford, Oxford University Press.

Freestone, I.C. (2015) The recycling and reuse of Roman glass: analytical approaches. *Journal of Glass Studies* 57, 29–40.

Gaggadis-Robin, V. and de Larquier, N. (eds) (2019) *La sculpture et ses remplois. Actes des IIes rencontres autour de la sculpture romaine, Arles 2016. L'atelier du sculpteur. Vol. 1.* Bordeaux, Ausonius Éditions.

Paynter, S. and Jackson, C.M. (2016) Re-used Roman rubbish: a thousand years of recycling glass. *Post-Classical Archaeologies* 6, 31–52.

Peña, J.T. (2007) *Roman Pottery in the Archaeological Record*. Cambridge, Cambridge University Press.

Pollard, A.M., Bray, P., Gosden, C., Wilson, A. and Hamerow, H. (2015) Characterising copper-based metals in Britain in the first millennium AD: a preliminary quantification of metal flow and recycling. *Antiquity* 89.345, 697–713. doi: 10.15184/aqy.2015.20.

Ponting, M.J. (2020) Recycling and Roman Silver Coinage. In C.N. Duckworth and A. Wilson (eds) *Recycling and Reuse in the Roman Economy*, 265–284. Oxford, Oxford University Press.

Sainsbury, V. (2019) *From Trash to Treasure. The Recycling of Glass in Roman and Early Medieval Period Britain*. Unpublished thesis, University of Oxford.

Sainsbury, V. and Liu, R. (eds) (2022) *Archaeometry Special Issue: Recycling* vol. 64, supplement 1.

Wilson, A.I. (2007) The metal supply of the Roman Empire. In E. Papi (ed.) *Supplying Rome and the Empire: The Proceedings of an International Seminar Held at Siena-Certosa di Pontignano on May 2-4, 2004, on Rome, the Provinces, Production and Distribution*, 109–125. Portsmouth, RI, Journal of Roman Archaeology.

Y Ng, D. and Swetnam-Burland, M. (2018) *Reuse and Renovation in Roman Material Culture: Functions, Aesthetics, Interpretations*. Cambridge, Cambridge University Press.

Chapter 1

The ragpicker's dream: notes on the continuous role of junk dealers in past urban economies from the Roman period onwards

Guido Furlan

Introduction

It has been proposed by Stuart Needham *et al.* (1989, 384) that effective recycling economies are by their nature invisible in archaeological terms. I largely agree with this; indeed, a combination of a wide array of sources, ranging from archaeometry to literary sources, is usually needed to identify several ancient reuse practices. This invisibility is even more evident when we try to investigate the people who made their living through reuse and recycling. In other words, the people behind the processes. When it comes to these individuals, the sources are even scarcer, and we are forced to combine these few sources with proxy data and analogies with socio-economic systems that are better known.

Although the data are far from being satisfactory, when considering circular economic activities and agents over a long period of time, an incredible example of continuity is represented by the activity of junk dealers. Junk dealers are known by several names: ragpickers, rag and bone men, *chiffonniers*, *straccivendoli*, *rigattieri*, *ferrovellers*, *ropavejeros* etc. Local nicknames exist in even larger numbers, and each of these names has a particular nuance. A specific definition of 'junk dealers' is somehow difficult, as these figures often blend with characters such as pedlars, hawkers, scavengers and garbage men, on the one hand, and with higher status figures such as proper traders, second-hand shopkeepers and even pawnbrokers, on the other hand. In other words, the group is largely heterogeneous in terms of social and economic status, and the basic guideline considered in this paper for the identification of 'junk dealers' is simply the activity of collecting and reselling damaged and/or discarded items of various kinds. The aim is to briefly trace the existence and role of this group of people in different periods, eventually attempting to understand their impact on the Roman economy.

The contemporary and the modern era

People who made their living roaming around with a bike and/or cart, collecting and reselling discarded items, particularly old clothes, furniture and metalware, were still popular in many European towns just a few decades ago, at least until the end of the 20th century. In Italy, these individuals were still to be found following the economic boom that occurred after the Second World War and before the growing awareness of the importance of circular economic models developed in the last decade. Junk dealers clearly managed to survive for a while in a highly consumeristic society with little or no awareness of the importance of large-scale recycling.

Garbage collectors and junk dealers still marginally survive in some Western economies, but they play a more substantial role in the so-called developing world (Wilson *et al.* 2009). In 2000, it was estimated that 2% of the population of Africa, Asia and Latin America, at the time, made their living by recovering materials from waste; in Cairo the population of refuse collectors included 30,000 people, the backbone of the waste management and recycling system (Medina 2000, 52–55).

In Western countries, junk collectors and dealers could draw on a well-established 19th-century tradition. Rag and bone men were popular throughout Europe, including France, Germany and Britain, and they were very common figures even in the United States (Strasser 1999; Zimring 2005). These individuals played a key role in the production of paper and cardboard, a role that ragpickers had played from at least the mid-medieval period (Cardinale 2018; Craig 2019). In cities such as New York and London, the life of rag and bone men is well known from several historical and iconographic sources, including photographic collections; in Paris, *les chiffonniers* were so popular that they even became literary characters. Therefore, we can base our knowledge of the lives of these individuals on solid ground (Compagnon 2017). For instance, there is clear evidence that during the 19th century this group included women and children, and that it was structured in a sort of pyramidal hierarchy with, from bottom to top, single collectors, intermediaries and dealers. We also know that part of the economic circuit of this trade was not monetised but based on simple barter (Strasser 1999).

The material, physical world of ragpickers, besides trash, consisted of backpacks, carts and wagons (Figs 1.1 and 1.2), as well as structures such as depots and exchange centres. It is noteworthy that many of these traits are present also in previous historical periods. A brief examination of the statute of the *Università dei rigattieri* (the guild of junk dealers) in 17th- and 18th-century Rome shows the very same traits just described for the 19th century (Travaglini 1992). The statute also provides further interesting elements, such as, obviously, the existence of a guild, a proper corporation regulated by internal norms and by law. The statute makes an explicit distinction between itinerant junk dealers and those who owned a proper deposit or a shop, implicitly assuming that the latter were richer than the former, and therefore confirming the social and economic heterogeneity of this group. The statute also mentions the establishment of low fees for guild membership for female junk dealers.

A further important point that emerges from an examination of this statute is the development of the role of junk dealers through time. The figure of the *rigattiere* does not remain the same through the decades: important new and remunerative economic functions developed over the course of these centuries. The roles of pawnbrokers and of valuers/appraisers, for example, are the most important, and they go hand in hand with changes in social status. It is also clear that there was explicit conflict between this and other guilds: in the case of Rome, particularly with the guild of goldsmiths. This friction arises from the economic role assigned to the guild of *rigattieri* by both the public authorities and the other guilds. This in turn suggests the importance of this circular economic circuit in the urban life.

In addition, we know from a variety of sources that different minorities, including the Jewish community, played an important role in this field. The activity of the Jewish community, often stereotyped and distorted by coeval sources (Todeschini 2018), is well attested throughout Italy and Europe, for instance, in Venice during

Fig. 1.1 The Rag Picker, 1911. Jean-François Raffaëlli (French, 1850–1924). Etching, drypoint and aquatint; The Cleveland Museum of Art, Gift in memory of Paul H. Oppman Sr. from his family. Accession number 1981.230. (Freely available on Internet Archive.).

Fig. 1.2 Woman and two dogs pulling wagon full of refuse. New York, 1870. Harper's Weekly, *May 7, 1870, p. 301. (Freely available at the Library of Congress. Digital ID: cph 3c06383 //hdl.loc.gov/loc.pnp/cph.3c06383).*

the 15th and 16th centuries (Ravid 1975). An understanding of these elements adds further complexity to the heterogeneous and palimpsestic world of early modern junk dealers.

The Middle Ages

The older statutes of the Roman guild of junk dealers go back as far as 1609, but similar regulations existed in other mid- and late and medieval cities (for the Netherlands, see Deceulaer 2008, esp. 15–16; for the city of Valencia, see García Marsilla 2016). Such regulations are well known in Florence, where one of the oldest statutes dates to 1296 (Sartini 1940). In this context, the main issue was creating a clear definition of the role of ragmen, particularly by fixing a clear boundary between them and the guild of textile producers. Ragpickers were allowed to collect and distribute rags and old clothes, for example, but they were forbidden from processing them in any way. This clear definition of the boundaries of their activity suggests, again, that disputes with other guilds occurred, due to the substantial economic interests at stake.

The examination of the figures of *rigattieri*, *cenciai* and *ferrovecchi* for the whole of Tuscany in the late medieval period, mainly thanks to the work of Alessia Meneghin (2015a; 2015b; 2016; 2020), demonstrates two other interesting points. Firstly, it is worth noting the presence of women, although few in number, involved in the market as second-hand dealers. Secondly, there was clearly a stratified structure related to the materials collected by different subgroups. *Rigattieri* dealt primarily, although not exclusively, with old clothes on a mid- to large scale, and they often had a proper store. They can be considered at the top of the hierarchy of junk dealers. Moreover, they were even occasionally at the head of structured activities, with employees assigned to street sales, provided with carts and donkeys. Metal scrap dealers (*ferrivecchi*) occupied a lower position both in economic and social terms, together with *cenciai* and *straccieri*, proper itinerant ragpickers, who collected and traded second-hand clothes and rags that were employed by the paper industry.

The figure of a scrap dealer from this period is well represented by Pietro di Lorenzo di Ser Lotto, a low status *rigattiere* in Volterra (Meneghin 2016, 19–20). About 70 years old, he lived in a small house that was provided with a store containing metal scraps, buffalo horns, gloomy leather garments, pork grease, saddles, old ironware, cookware, other goods of various nature and gloomy low-quality shoes. Figures such as Pietro di Lorenzo are attested in other European countries throughout the medieval period, indicating not only the continuity of the activity of junk collecting and reselling through time but also its widespread spatial distribution; for instance the *post mortem* inventories of two Valencian *pellers* (pawnbrokers and dealers in second-hand clothes and textiles), redacted in 1344 and in 1415, vividly depict the goods and the activities of two higher status traders (García Marsilla 2016, 354–356).

What has been sketched clearly represents just a very brief and incomplete overview; much more could be added, but the point is that reconstructing the

Table 1.1 Excerpt from the list of jobs in the medieval Islamic world published in Shatzmiller 1994.

Job name	*Translation*	*Place*	*Centuries*
anmāṭ	merchant of used and new items	Egypt	14th
anmāṭī	seller of rugs	Egypt	14th
basṭānī	hawker, peddler	Iraq, Egypt, Syria	9th, 10th, 11th
bayyāʿ sakaṭ	junk dealer, rag man, seller of baskets	Egypt	15th
busuṭī	rag seller	Egypt	14th
mayyār	peddler	North Africa	14th
rabb al-maḳāʿid	peddler	Egypt	14th
ṣāḥib al ḵulḳān	rag merchant	Iraq	8th, 9th, 10th, 11th
ṣaḳaṭī	trash, junk dealer, seller of old household goods	Iraq, Egypt, Syria	9th, 10th, 11th, 12th, 14th, 15th
ṣaḳḳāṭ	junk dealer	Spain	11th
ṭawwāf	house to house peddler, storyteller, wandering merchant	Spain, Egypt	13th, 14th

activity of junk dealers in the mid-Middle Ages can be achieved straightforwardly and consistently.

Tracing the activity of junk dealers and rag and bone men in the early Middle Ages is more difficult. To partially overcome the lack of data and sources for these periods, it is necessary to look at to the whole Mediterranean basin. A list of jobs in the medieval Islamic world, published by Maya Shatzmiller (1994), includes activities related to several circular economic practices; some of these jobs are clearly related to the collection and trade of used and/or discarded items (Table 1.1). While the figure of a hawker can be considered generic, the descriptions of rag merchants and proper junk dealers are very specific. Therefore, there seems to be no doubt about the existence of people carrying out a set of activities extremely similar to those that existed in medieval Italy. Even more interestingly, these sources allow us to identify junk dealers as far back as the 8th or 9th centuries AD, particularly in regions which had still been part of the Byzantine Empire during the first decades of the 7th century AD.

The Roman period

Filling the gap between about the 8th century AD and antiquity is extremely difficult, as already stressed by previous scholarship (Cracco Ruggini 1971; Mor 1971; Greci 1995; Galletti 2017). The debate about the continuity or discontinuity between Byzantine *scholae*, Langobard and Carolingian *officia* and *ministeria*, and Roman *collegia* turns out to be, in our case, irrelevant, as there is no evidence for the existence of guilds of second-hand dealers in this period. However, if we move chronologically backwards into the Roman world, we immediately find the well-known *collegium* of *centonarii* (see

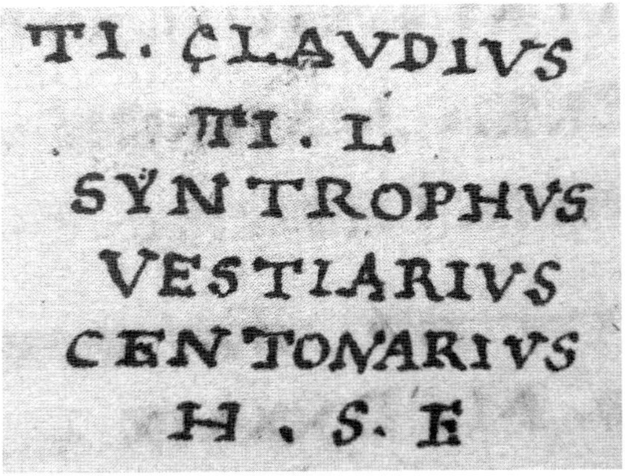

Fig. 1.3 The inscription of Ti. Claudius Syntrophus, in the manuscript of Gian Domenico Bertoli (BERTOLI ms. 7493, f 39r, n. DCCXLIV), attesting the existence of a vestiarius centonarius *in Aquileia.*

Gleba and Busana in this volume); the *collegium* was certainly still active at the beginning of the 5th century AD, as it is mentioned three times in the *Codex Theodosianus* and attested by an inscription dated to AD 367 (Liu 2009, 279–294).

The debate on the role of *collegia centonaria* (*centones* were rags) has been active for decades. One hypothesis identifies *centonarii* as firefighters. While this is possible, particularly for the period following the 2nd century AD, there seems to now be consensus that at least part of the *collegia* included textile producers and/or dealers of some sort (Liu 2009; Wild 2020). The role of the *centonarii* has been confirmed by a 1st-century AD inscription from Aquileia mentioning a *vestiarius centonarius* (Fig. 1.3), and the authenticity of the discovery has been convincingly demonstrated by Alfredo Buonopane (2003). The exact social composition of this group is still unclear, and it may have been, again, anything but homogeneous. Likewise, we are not informed in detail about the activities carried out by the *centonarii*. Nonetheless, a few points can be stated with some confidence: *centonarii* dealt with *centones* (small, cheap pieces of fabric, usually, but not exclusively, made of wool or felt); *centones* were used by both the military and civilians, particularly the poor; *centones* were used for patchwork to produce new garments; and *centones* were likely largely second-hand rags. Lastly, it is likely that the activity of the *centonarii* included, or was related to, the sale of second-hand clothes; this, in turn may have involved individuals such as pawnbrokers (Liu 2009, 72). All these aspects, besides vividly recalling people and activities known for later periods, also suggest that the *centonarii* played a vital role in the circular economy of Roman textiles. The presence in the Roman society of large numbers of people with little spending power indicates that the market for clothes made of reused textiles should not be considered as secondary, as it has been demonstrated that domestic production alone could not have met the whole demand (Liu 2009, 74).

Individuals who were not necessarily organised in the *collegium* certainly performed the physical picking and collection of old clothes or rags that were to be resold or reused for low-quality garments. Again, it seems possible to catch a glimpse of an organisation like the one observed in medieval and modern towns centuries later, namely a hierarchy with low level *ambulatores* and higher-level dealers, probably not very dissimilar to the most successful among the cited medieval and post medieval *rigattieri*. The existence of a *collegium*, by itself, and the possible involvement of pawnbrokers represent elements of noteworthy similarity between periods.

Although the evidence for the activity of the *centonarii* is still patchy, their role in second-hand trade appears more convincing when framed in the overall set of reuse activities documented in Roman towns. Discarded clothes and textiles were not the only collected items to be reemployed. The recycling of glass is particularly well documented by archaeometric analyses and by the occasional discovery of cullet stored before it was remelted (the literature on the topic is extremely vast; see, in particular, Keller 2005; Freestone 2015; Duckworth 2020).

The available literary sources, although not generous, are clear about the existence of hucksters wandering in Roman cities, collecting broken glass and exchanging it for sulphur sticks (something, most probably, similar to modern matches). Two passages in Martial (1.41 and 10.3), one in Juvenal (5.48) and one in Statius (*Silv.* 1.6) clearly attest to the existence of this practice (and, more importantly, of these people) in Rome during the second half of the 1st century AD. Although these figures are cited for literary reasons, there is no doubt that the three authors drew from common experience (Leon 1941; Whitehouse 1999).

In this regard, it is important to stress a few aspects of these passages. Firstly, the nonchalance showed by the three authors when mentioning the activity of exchanging sulphur sticks for cullet suggests that this activity was well known to the reader, who could therefore immediately understand the meaning of the passages. This fact demonstrates that this activity was not occasional or uncommon. Secondly, the people depicted seem to belong to the lower classes of the urban community. In one of the passages by Martial (1.41) the collector of broken glass is qualified as a *transtiberinus ambulator*. The region beyond the River Tiber hosted unpleasant activities, such as tanning, and it was occupied by the lower classes, as well as a Jewish community (Howell 1980, 193). Lastly, it is important to note that the exchange depicted in the literary sources is not monetised; it is a form of barter, a practice, as we have seen, that persisted through the ages as part of the so-called informal economy.

We cannot know if the collectors of cullet were specifically part of the glass industry or if they also collected other reusable items door to door. However, it is certain that other collectors and retailers of second-hand goods existed: this is supported by a passage of the *Noctes Atticae* (3, 14, 10), published by Aulus Gellius in the second half of the 2nd century AD but largely drawing from previous literature. Gellius, focusing on a very specific grammatical issue, introduces the figure of a junk dealer mentioned by the 2nd-century BC poet Lucilius; this character, although probably fictional, is most likely drawn from real figures somehow typical of the streets of Rome (see Holleran 2012). The junk dealer mentioned in the text is yelling to sell his poor merchandise, namely part of a strigil and half a shoe (the association of metal scrap and damaged shoes interestingly recalls the merchandise in the store of Pietro di Lorenzo). Most importantly, the text provides the term for the Roman junk dealer (or, at least, one of the possible names that he or she could take), *scrutarius*. The name, derived from the verb *scruto*, is extremely evocative as it perfectly describes the activity of rummaging through refuse to sort out what could be reused and what could not. In other words, it indicates the profitable skill of seeking out valuable items.

This passage shares many traits with an excerpt from an *epistula* by Horace (I.7.64–66) that mentions the sale of *scruta* to common people in the streets, but in this second case the reference to proper junk (meaning something which was to be trashed) is less explicit. The passage in Gellius also casts invaluable light on the meaning of an inscription recovered in Kos and published by Amedeo Maiuri in 1925 (Maiuri 1925, 168 n. 466). The bilingual inscription, carved on a marble base and dedicated to Augustus, mentions the existence of a group of *scrutarei,* who were organised into a proper association with a guild leader. The Greek term for *scrutarei,* γρυτοπῶλαι, also mentioned in the inscription, led Maiuri to interpret these traders as 'rivenditori di profumi ed oggetti di toletta femminile'; however, the activity described by Gellius and carried on by a *scrutarius* clearly points to a more general interpretation as junk dealers. Curiously, Maiuri himself, in relation to the retrieval of this inscription, stressed the existence in Kos of a rich Jewish community, which is attested from the 3rd century BC onwards (also mentioned in Flavius Josephus 14, 7, 2).

The activity of auctioneers (*praecones*; see Rauh 1989; Donadio 2007) overlaps with that of junk dealers and with that of pawnbrokers; it also blends to some extent with that of proper bankers and moneychangers (*argentarii*), both of whom played a fundamental role in disposing of substantial fortunes, such as inheritances. This group included both public heralds and private auctioneers of different status. Their activity embraced the sale of war booty, the disposal of properties confiscated by the state, the sale of the former properties of debtors and even the periodic distribution of wares and foodstuff in the city. However, they were also middlemen for anyone who wanted to earn money getting rid of used items.

The sale of second-hand goods on behalf of third parties is well depicted in a satire by Juvenal (7.9–11): cupboards, tables, chests, jugs and books are listed as items that could be resold to make a living thanks to the commission applied. Whereas the auction of major assets (including land plots and slaves) involved the activity of *argentarii* and was well coded, smaller and occasional transactions, such as the one depicted by Juvenal, seem to resemble more closely the activity of proper junk dealers and pawnbrokers. The disposal of inherited items and, more generally, of second-hand goods is mentioned also in the *Digesta Iustiniani* (*e.g.*, 18.I.45; 33.VII.18.13); niche activities, such as the trade of second-hand books or whole libraries, are also known (Starr 1990), but an economic appreciation of these activities is extremely difficult.

It has already been noted that in contemporary developing countries and during the modern and medieval periods in Europe, the role of dealers in second-hand goods is far from insignificant. However, there is less known about the economic impact of junk collectors in the Roman economy. Unfortunately, the better-known corporation of these individuals, the *centonarii*, dealt with organic material which is seldom preserved in the archaeological record. This generally prevents us from using archaeological data as a reliable quantitative proxy for the economic impact of ragpicking. One extraordinary exception is presented in this volume by Margarita

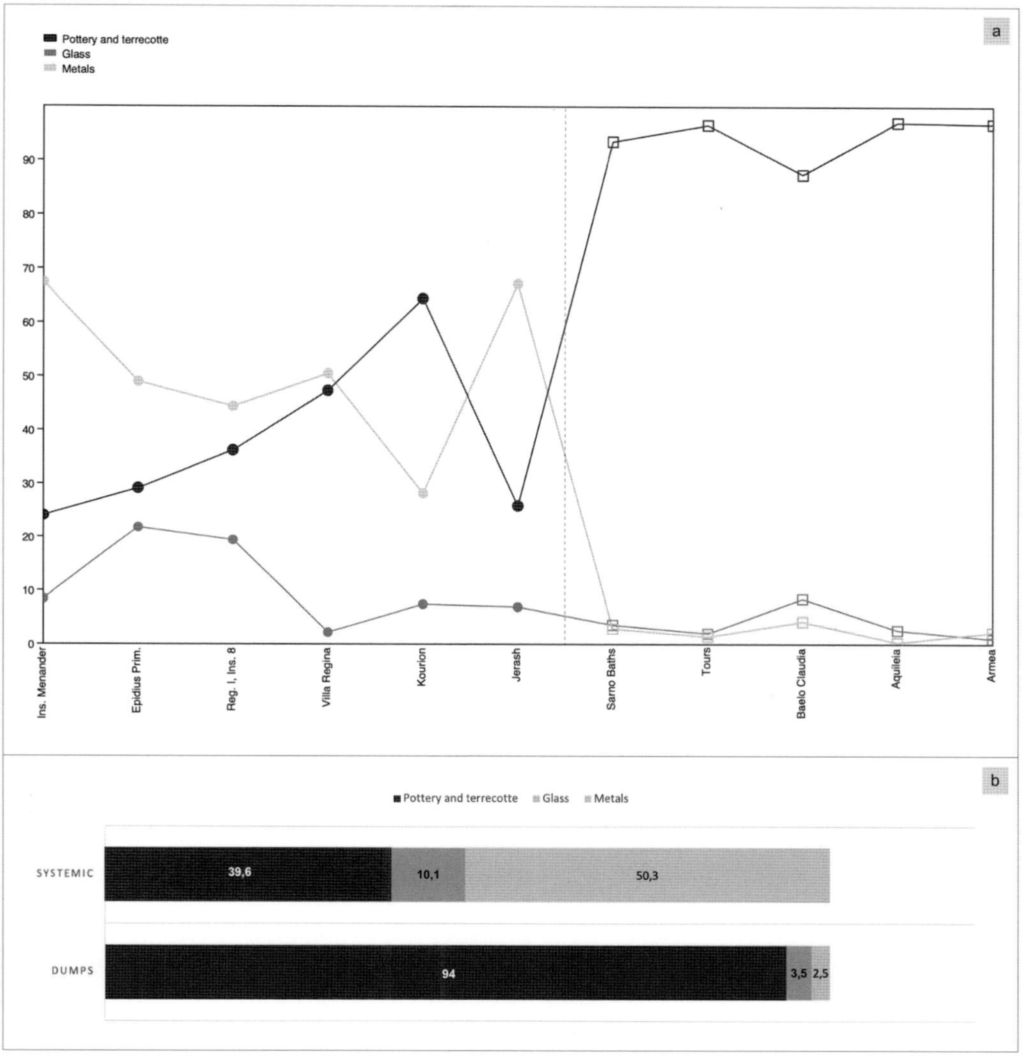

Fig. 1.4 a: The varying percentages of glass, metal and ceramics in the contexts examined in Furlan and Andreatta forthcoming (dot=primary context in situ, *square=dumping site). b: The overall difference in the percentage of glass, metal, and ceramics between primary contexts* in situ *(mirroring the systemic context as much as possible) and dumping sites.*

Gleba and Maria Stella Busana. However, we can try to quantify the impact of the routine collection of reusable items using other proxies, such as glass and metal finds.

Recently a comparison was made between assemblages contained in well-sealed, primary domestic and commercial contexts (*in situ*, with low disturbance) and assemblages recovered in *extra moenia* communal dumps (Furlan and Andreatta 2023). This can be roughly translated into comparing what was in use in a given Roman town in a certain period with what was discarded. In the primary contexts, there were high percentages of glass and metal items (around 10% and 50%, respectively, with 40% ceramic objects; building and organic materials were not counted). In contrast, metals and glass were almost completely absent from dumping sites (Fig. 1.4). The many factors that came into play to shape these two different types of assemblages have been considered, and it has been shown that effective collection and recycling mechanisms were the main cause of the near absence of metal and glass observed in the dumps. Almost nothing was wasted and the absence of certain materials in rubbish deposits is a proxy for the intense reuse activities related to these materials.

This clearly implies not only the existence of well-structured and routine practices but also the presence of actors who physically provided for the collection, selection and distribution of what was discarded daily and worth being reused. These agents can be clearly identified as the range of individuals with different social statuses and different specific jobs that have been summarily sketched above from the available written sources and that find a very good match with the figures (*rigattieri*, *ferrivecchi*, *chifonniers* etc.) known from later periods. At different stages of the so-called urban waste stream, these people picked out materials that then re-entered the economic circuit, therefore acting as engines for the local circular economy. The impact of this range of activities (and of these people) has been used by Emilio Rodríguez-Almeida (2000) to argue that Rome was a self-cleaning city. E. Marianne Stern (1999), examining the prices of raw glass and finished glass objects in the Edict of Diocletian, concluded that a glassblower's activity could barely survive without limiting waste and relying heavily on recycling. Those *ambulatores* providing glass scrap played a substantial role in a non-secondary sector of the Roman economy.

Conclusions

We can conclude that what we could call a circular approach to materials and resources was far from unknown in Roman cities; it made the local economy more sustainable and effective, and it reduced the dependency on raw materials (Furlan 2023). Junk dealers, auctioneers, collectors, peddlers, *centonarii* and, most probably, even dump miners were at the centre of this circuit. As anticipated, the world of Roman junk dealers seems to be perfectly aligned and comparable with what has emerged, increasingly clearly, from later periods and up to the present day. What comes to light from this examination is indeed a considerable form of continuity (see Downs and Medina 2000), where the Roman world is part of a shared trend. Throughout the last two millennia, collecting and recycling scrap, in the framework of a more or less sophisticated structure, were activities well rooted in urban (and rural) communities. They shaped the economic and social life of groups who were often marginalised, and they had an important economic function, reducing waste and optimising resources. From the point of view of archaeological research, reuse activities and their agents markedly contributed to shaping the material record of the very systems under investigation.

The activity of collectors, junk dealers and ragpickers shows continuity through time in various forms:

1. There is considerable continuity in the materials involved, with a pivotal role played by textiles and metal scrap, although clearly new recyclable materials were introduced through time (chiefly paper and cardboard).
2. There seems to be persistence in social structures and organisation, with richer dealers often organised in proper guilds and single, poorer collectors (structured heterogeneity). The inclusion in this circuit of women and most probably children (possibly the most unknown economic agents in archaeological and historical research) also seems to be a trait of continuity, although, so far, it cannot be demonstrated for the Roman period.
3. The role of collectors and junk dealers between the so-called formal economy and the informal economy, which was made of small unregulated trade, often involving barter, also seems to be a *trait d'union* through the centuries.

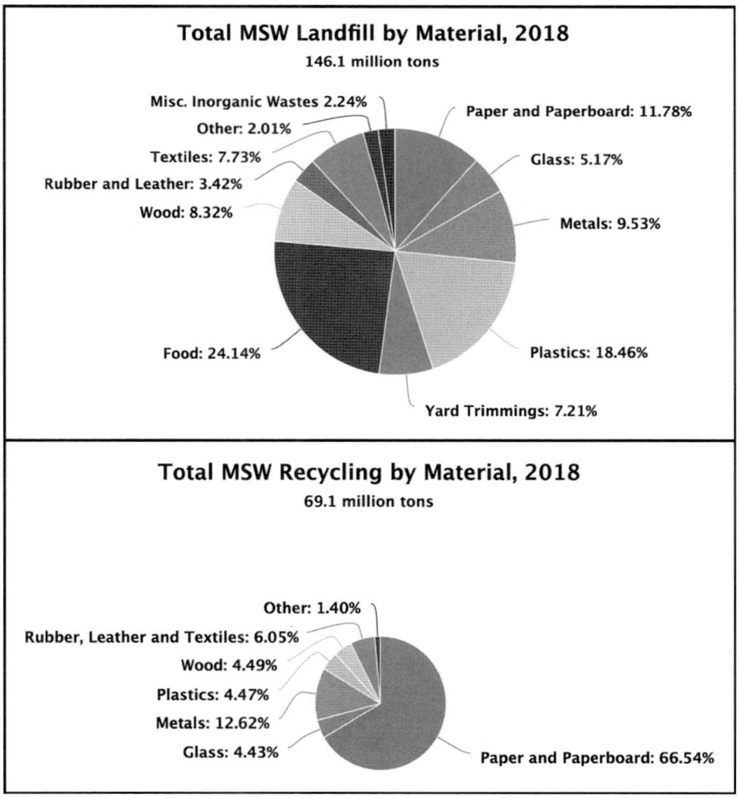

Fig. 1.5 American municipal solid waste according to the United States Environmental Protection Agency: landfill composition (top) and recycling (bottom) by material, 2018. (https://www.epa.gov/facts-and-figures-about-materials-waste-and-recycling/national-overview-facts-and-figures-materials).

4. The quantitative impact of these individuals on the local economy seems to have been relevant in all periods, as shown by archaeological data for the Roman period and indicated by juridical and historical sources for later periods (*e.g.*, the frictions with other guilds). A better appreciation of the fluctuations of this impact through time is certainly an aspect worthy of future investigation.

Overall, one has the impression that, in a world that changed immensely over the last two millennia, junk dealers and their activities, particularly at the humblest level, remained relatively unchanged. The considerable exception to this general trend is represented by the Western world in the most recent period, which is apparently only just now rediscovering the benefits of a widespread circular approach to the economy and learning to practice it in new and more sophisticated forms. However, the limits of this new awareness are still clearly visible: it is notable that glass and metal recycling in the current garbage disposal system in the United States would have been no match in terms of efficiency for the ancient Roman system (Fig. 1.5; note the presence of abundant organic refuse, which is obviously not recovered in most archaeological dumps; see also Rathje and Murphy 1992; Humes 2012).

In conclusion, while there is limited evidence in the earliest periods to write a proper social history of junk dealers (and there may never be, given the amount of available data), the impact of these individuals can be perceived, admittedly still feebly. This variegated community held continued value to the circular economy at least over the course of the last two millennia. If it is true that recycling economies are by their nature invisible in archaeological terms, then it must be admitted that junk collectors and dealers did a very good job throughout the centuries.

Acknowledgements

This research was funded by the Danish National Research Foundation under the grant DNRF119 – Centre of Excellence for Urban Network Evolutions (UrbNet). I wish to thank Prof. A. Buonopane for the fruitful discussion of several aspects of the topic.

Bibliography

Primary sources

Digesta Iustiniani. droitromain.univ-grenoble-alpes.fr

Flavius Josephus. *Jewish Antiquities*. Volume VI: Books 14–15. Translated by Ralph Marcus, Allen Wikgren. Loeb Classical Library 489. Cambridge, MA, Harvard University Press, 1943.

Gellius. *Noctes Atticae*. Volume I: Books 1–5. Translated by J. C. Rolfe. Loeb Classical Library 195. Cambridge, MA, Harvard University Press, 1927.

Horace. *Epistulae*. Satires. Epistles. The Art of Poetry. Translated by H. Rushton Fairclough. Loeb Classical Library 194. Cambridge, MA, Harvard University Press, 1926.

Juvenal. *Saturae*. Juvenal and Persius. Edited and translated by Susanna Morton Braund. Loeb Classical Library 91. Cambridge, MA, Harvard University Press, 2004.

Martial. *Epigrams*. Volume I: Spectacles, Books 1–5. Edited and translated by D. R. Shackleton Bailey. Loeb Classical Library 94. Cambridge, MA, Harvard University Press, 1993. Volume II: Books

6–10. Edited and translated by D. R. Shackleton Bailey. Loeb Classical Library 95. Cambridge, MA, Harvard University Press, 1993.
Statius. *Silvae*. Edited and translated by D. R. Shackleton Bailey. Revised by Christopher A. Parrott. Loeb Classical Library 206. Cambridge, MA, Harvard University Press, 2015.

Secondary sources

Buonopane, A. (2003) Un *vestiarius centonarius* ad Aquileia: sulla genuinità di CIL, V 50. *Aquileia Nostra* LXXIV, 301–314.
Cardinale, C. (2018) *A Penny for Your Rags: Rag Pickers and the Paper Industry in the Later 19th Century*. Unpublished Thesis, Buffalo University.
Compagnon, A. (2017) *Les Chiffonniers de Paris*. Paris, Éditions Gallimard.
Cracco Ruggini, L. (1971) Le associazioni professionali nel mondo romano-bizantino. In Centro italiano di studi sull'alto medioevo (ed.) *Artigianato e tecnica nella società dell'alto medioevo occidentale. Settimane di studio del Centro italiano di studi sull'alto medioevo*, 59–193. Spoleto, Centro italiano di studi sull'alto medioevo.
Craig, H. (2019) Rags, ragpickers, and early modern papermaking. *Literature Compass* 16. 5, 111.
Deceulaer, H. (2008) Second-hand dealers in the Early Modern Low Countries. In L. Fontaine (ed.) *Alternative Exchanges. Second-Hand Circulations from the Sixteenth Century to the Present*, 13–42. New York/Oxford, Berghahn Books.
Donadio, N. (2007) Le 'auctiones' private all'epoca di Plauto. Consuetudini, regole, pratiche nelle vendite all'asta nel mondo romano e loro tracce nella 'palliata' Latina. In E. Cantarella and L. Gagliardi (eds) *Diritto e teatro in Grecia e a Roma*, 117–197. Milano, Led.
Downs, M. and Medina, M. (2000) A short history of scavenging. *Comparative Civilization Review* 42, 23–45.
Duckworth, C.N. (2020) Seeking the invisible. New approaches to Roman glass recycling. In C.N. Duckworth and A. Wilson (eds) *Recycling and Reuse in the Roman Economy*, 301–356. Oxford, Oxford University Press.
Freestone, I.C. (2015) The recycling and reuse of Roman glass: analytical approaches. *Journal of Glass Studies* 57, 29–40.
Furlan, G. (2023) Aspects of circular economy on an urban and inter-urban scale in the Roman age: towards a more holistic model. *Thiasos* 12, 341–366.
Furlan, G. and Andreatta, C. (2023) Waste nothing. An archaeological approach to the impact of glass and metal recycling in Roman imperial towns. *European Journal of Archaeology*, 1–19.
Galletti, F. (2017) Le società delle arti a Bologna e i loro statuti: un bilancio storiografico. *I quaderni del m.æ.s.* XV, 69–93.
García Marsilla, J.V. (2016) Expertos de lo usado. Pellers, ferrovellers y corredors de coll en la Valencia medieval. In L. Feller and A. Rodríguez (eds) *Expertise et valeur des choses au Moyen Âge. II. Savoirs, écritures, pratiques*, 343–358. Madrid, Casa de Velázquez.
Greci, R. (1995) Le corporazioni. Associazioni di mestiere nell'Italia del Medioevo. *Storia e Dossier* 99, 71–97.
Holleran, C. (2012) *Shopping in Ancient Rome. The Retail Trade in the Late Republic and the Principate*. Oxford, Oxford University Press.
Howell, P. (1980) *A Commentary on Book One of the Epigrams of Martial*. London, The Athlone Press.
Humes, E. (2012) *Garbology. Our Dirty Love Affair with Trash*. London, Penguin.
Keller, D. (2005) Social and economic aspects of glass recycling. In J. Bruhn, B. Croxford and D. Grigoropoulos (eds) *TRAC 2004: Proceedings of the Fourteenth Annual Theoretical Roman Archaeology Conference, Durham 2004*, 65–78. Oxford, Oxbow Books.
Leon, H.J. (1941) Sulphur for broken glass (Martial 1.41.3-5). *Transactions and Proceedings of the American Philological Association* 72, 233–236.

Liu, J. (2009) Collegia Centonariorum: *The Guilds of Textile Dealers in the Roman West*. Columbia Studies in the Classical Tradition 34. Leiden, Brill.

Maiuri, A. (1925) *Nuova silloge epigrafica di Rodi e Cos*. Firenze, Le Monnier.

Medina, M. (2000) Scavenger cooperatives in Asia and Latin America. *Resources, Conservation and Recycling* 31, 51–69.

Meneghin, A. (2015a) The trade of second-hand clothing in fifteenth-century Florence: organisation, conflicts, and trends. In G. Nigro (ed.) *Il commercio al minuto. Domanda e offerta tra economia formale e informale. Secc. XIII–XVIII*, 319–336. Firenze, Firenze University Press.

Meneghin, A. (2015b) The second-hand clothing business of Taddeo di Chello (1390–1408). Social relations, credit, and the structures of consumption in late medieval Prato. *Quaderni Storici* 2, 523–550.

Meneghin, A. (2016) Rigattieri, cenciai e ferrovecchi dello stato territoriale fiorentino: un'indagine preliminare, 1428–1429. *Ricerche Storiche* XLVI. 3, 5–28.

Meneghin, A. (2020) *The Social Fabric of Fifteenth-Century Florence. Identities and Change in the World of Second-Hand Dealers*. New York/London, Routledge.

Mor, C.G. (1971) Gli artigiani nell'alto medioevo (con particolare riguardo ai riflessi giuspubblicistici). In Centro italiano di studi sull'alto medioevo (ed.) *Artigianato e tecnica nella società dell'alto medioevo occidentale. Settimane di studio del Centro italiano di studi sull'alto medioevo*, 195–213. Spoleto, Centro italiano di studi sull'alto medioevo.

Needham, S.P., Leese, M.N., Hook, D.R. and Hughes, M.J. (1989) Developments in the early Bronze Age metallurgy of southern Britain. *World Archaeology* 20.3, 383–402.

Rathje, W. and Murphy, C. (1992) *Rubbish! The Archaeology of Garbage*. New York, Harper Collins.

Rauh, N.K. (1989) Auctioneers and the Roman Economy. *Historia: Zeitschrift für Alte Geschichte* 38.4, 451–471.

Ravid, B. (1975) The legal status of the Jewish merchants of Venice, 1541–1638. *The Journal of Economic History* 35.1, 274–279.

Rodríguez-Almeida, E. (2000) Roma, una città self-cleaning? In X.D. Raventós and J.A. Remolà (eds) Sordes Urbis: *la eliminación de resduos en la ciudad romana*, 123–127. Roma, L'Erma di Bretschneider.

Sartini, F. (ed.) (1940) *Statuti dell'arte dei rigattieri e linaioli di Firenze: (1296–1340)*. Firenze, Le Monnier.

Shatzmiller, M. (1994) *Labour in the Medieval Islamic World*. Leiden/New York/Köln, Brill.

Starr, R.J. (1990) The used-book trade in the Roman world. *Phoenix* 44.2, 148–157.

Stern, E.M. (1999) Roman glassblowing in a cultural context. *American Journal of Archaeology* 103.3, 441–484.

Strasser, S. (1999) *Waste and Want. A social History of Trash*. New York, Holt Paperbacks.

Todeschini, G. (2018) Jewish usurers, blood libel, and the second-hand economy. The medieval origins of a stereotype (from the thirteenth to the fifteenth century). In J. Adams and C. Heß (eds) *The Medieval Roots of Antisemitism. Continuities and Discontinuities from the Middle Ages to the Present*, 341–351. New York, Routledge.

Travaglini, C.M. (1992) Rigattieri e società romana nel Settecento. *Quaderni storici* 27.80.2, 415–448.

Whitehouse, D. (1999) Glass in the Epigrams of Martial. *Journal of Glass Studies* 41, 73–81.

Wild, J.P. (2020) The reuse of textiles in the Roman world. In C.N. Duckworth and A. Wilson (eds) *Recycling and Reuse in the Roman Economy*, 61–88. Oxford, Oxford University Press.

Wilson, D.C., Araba, A.O., Chinwah, K. and Cheeseman, C.R. (2009) Building recycling rates through the informal sector. *Waste Management* 29, 629–635.

Zimring, C.A. (2005) *Cash for Your Trash. Scrap Recycling in America*. New Brunswick, Rutgers University Press.

Chapter 2

Laws, letters and graves: the organisation of scavenging in the early medieval period

Irene Bavuso

Introduction

This contribution seeks to investigate the agents and the organisation of collection practices during the first centuries of the early medieval period in western Europe, with examples from Italy and England; in particular, it will concentrate on scavenging from cemeteries and burial grounds. Archaeological and historical research is continuing to highlight the ubiquity of practices of reuse during the early medieval period, and this surge of interest in forms of circularity has shown that they were widespread and effectively intrinsic to the economy well before the Middle Ages. Thus, recycling and reuse are no longer considered straightforward evidence for economic decline (Duckworth and Wilson 2020; Bavuso *et al.* 2023, for further references). Of course, the perceptions and motivations of the people who decided to reuse a material or an object could and did vary, in both space and time, so that what is valid for a flourishing Roman town likely will not apply to 5th-century England. The question of the validity of taking reuse as an indicator of economic vitality and decline at the beginning of the medieval period is a highly complex one, and reassessing the evidence in this perspective is beyond the scope of the present contribution. What this chapter tries to do is to complicate the picture, indicating possible avenues to understand better the socio-economic dynamics behind reuse – in particular, I will consider the coexistence of loose horizontal organisations and top-down ones, which involved social hierarchies, as well as the moral and literary reflections on and justifications of these acts.

Research on both written sources and archaeology strongly emphasises that we can now reach a much more complex understanding of these dynamics. As Line Van Wersch and Alexis Wilkin highlight in their contribution to the present volume and in their previous investigation of glass and glass recycling in early medieval Francia, important questions concern the organisation of workshops and recycling practices, and the nature and extension of circuits of distribution for reused materials within and across territories (Van Wersch and Wilkin, Chapter 4; Van Wersch *et al.* 2019; Wilkin 2019). Following a comparable line of inquiry, recent work on the aftermath of the Roman *villae* in western Europe by Beth Munro (2020) has indicated that the collection and distribution of scavenged materials may have been organised and conducted on a vast scale: scavenged materials from *villae* could travel quite widely, using, for instance, the river systems; this is evidence of a good level of organisation. So, we do not see only disorganised and uncoordinated practices. For the early medieval period, a key and much-cited case is the glass workshop at San Vincenzo al Volturno, in southern Italy, which illuminates very well the relationship between early medieval production at a large and high level, and the organised scavenging of a Roman site (Marazzi and D'Angelo 2009; Hodges *et al.* 2011; Schibille and Freestone 2013). More recently, the excavation conducted on the *emporium* at Ribe in Denmark has clearly demonstrated the great potential of investigating the long-distance travel of reused materials – glass, in this case (Barfod *et al.* 2022). These investigations, though on quite different areas and topics, show well that the issue of how collecting was organised has now entered historical and archaeological discourses.

The present contribution asks whether similar systematic practices may be detected in the case of the scavenging of cemeteries, and at what level this was managed; as will be discussed presently, both written sources and archaeology permit us to detect organisation, although its precise articulation remains, in many cases, difficult to understand. Given the fragmentary and largely anecdotal state of the sources, this contribution will stretch from Ostrogothic Italy to early post-Roman England. Although such a vast area witnessed profound differences at the socio-economic level, as well as in the organisation of political structures and the general continuity from Late Antiquity, the ubiquity of the practice in Europe makes it possible to include different territories in this investigation.

The first element to be acknowledged is that, even if we do possess a large number of mentions of tomb violations, who the people involved in collection practices were and whether and how these activities were organised remains unclear from the written sources of the earliest medieval period. Nonetheless, the existing sources are enough to allow us to explore the coexistence of various levels of organisation, the strong differences between the norms regulated by the laws and the reality of practices, and the moral justifications that were deployed to pursue violations of funerary grounds. The latter point needs to be stressed: cemeteries are radically different sites than public buildings or the above-mentioned *villae*. They carried profoundly different meanings for their communities, and this difference might in some cases be felt by the communities even after the places have stopped to be used. As a result, scavenging from cemeteries was often seen as an act of desecration in the written sources, and as such it was amply condemned by the laws. In contrast, scavenging from the ruins of *villae* was not a morally reproachable act. This difference may have important implications, as the hypothesis of a fully organised and top-down scavenging practice in cemeteries would clash with how this act was perceived.

Desecration and punishments

> I never insulted the dead or used tomb-stones for building, I swear by justice and the dead. Therefore bring no more iron to attack my stones, or if thou dost, let it fall on thy own head. It is I, Martinianus, who request this. If there be any gratitude for my glory, let my tomb remain for ever (AP 8.117. Trans. Paton, revised Tueller 2014).[1]

This epigram, included in Book 8 of the *Palatine Anthology*, is part of a group of poems against tomb violation by Gregory of Nazianzus (*c.* AD 329–390), and it is a fine example of the motif of maledictions and retaliations against grave-robbers that is widespread in late-antique funerary poetry (Floridi 2013; Moroni 2013; Goldhill and Greensmith 2020 on the *Palatine Anthology*, Book 8). In the same group of epigrams, the 4th-century theologian also makes the dead Martinianus address the violators and threaten them with his still-standing power:

> Away, far away! It is an evil exploit ye attempt, heaving up the stones of my tomb. Away! I am Martinianus. The living I benefited and here dead I have no little power. (AP 8.112. Trans. Paton, revised Tueller 2014).[2]

There are plenty of references to tomb violations in legal sources across the Mediterranean and Continental Europe from Antiquity and the Middle Ages that have been amply discussed (*e.g.*, Effros 2002; Rebillard 2003; Lafferty 2014; La Rocca and Tantillo 2017); here it is worth noting only some of the late-antique sources for the explicit connections they show between the scavenging of burial grounds and reuse practices, as well as between violations and the idea of contamination. For the 4th century, the *Codex Theodosianus* contains an edict of Constantius II, promulgated in AD 340, stating that the person violating a tomb – by demolishing it, in this case – is to be sentenced to forced labour in the mines or to prison. The stress of the edict may be on materials reused in buildings, which is generally the main interest of these laws.

> Emperor Constantius Augustus to Titianus, Prefect of the City.
>
> If any person should be apprehended in the act of demolishing a tomb, and if they should do this without the knowledge of their master, they shall be sentenced to the mines. If, however, they should be urged by the authority and command of their master, they shall be punished by the exile of relegation. If by chance anything taken from the tombs and carried to their house or villa should be discovered after the issuance of this law, the villa or house or any other building shall be vindicated to the account of the fisc (Trans. Pharr 1952, modified).[3]

A later edict by the same Constantius II, promulgated in the 350s, also clarifies that material taken from tombs (stones, marbles, or columns) could be used for building purposes or sold. Moreover, it states the reasons why violation was a crime: it was both the spoliation of tombs per se and the contamination of the living that were condemned.

> The same Augustus to the People.
>
> Those persons who violate the habitations of the shades, the homes, so to speak, of the dead, appear to perpetrate a twofold crime. For they both despoil the buried dead by the destruction of the tombs, and they contaminate the living by the use of this material in building. If any person, therefore, should take away from a tomb, stones or marbles or columns or any other material for the purpose of building or if he should do this for the purpose of selling such material, he shall be compelled to pay ten pounds of gold to the fisc [...] (Pharr 1952).[4]

The sanctity of tombs and the idea that their violation constituted a sacrilege are further expressed in another edict, this one by Julian in AD 363.

> Emperor Julian Augustus to the People.
>
> Criminal audacity extends to the ashes of the dead and their consecrated mounds, although our ancestors always considered it the next thing to sacrilege even to move a stone from such places or to disturb the earth or to tear up the sod. But some even take away from the tombs ornaments for their dining rooms and porticoes. We consider the interests of such criminals first, that they may not fall into sin by defiling the sanctity of tombs, and We prohibit such deeds, restraining them by the penalty which avenges the spirits of the dead (Pharr 1952, modified).[5]

Tomb violations continued to be sanctioned in later legislation and remained something that continued to elicit reflection, as the written sources show, during the early medieval period. In the *Lex Salica*, whose origins seem to be best ascribed to the late 5th century, the thief who robs a dead body is punished harder if the corpse was already buried than if it was an unburied one (MGH LL nat. Germ. 4.1, 14.9–10. See Effros 2002; Lafferty 2014 for later recensions of the code; see Ubl 2017 on the dating of the *Lex Salica*). The fact that tomb violation was reiterated in legal sources corroborates the idea that it was a widely common phenomenon – both written sources and archaeology do provide plenty of evidence for scavenging of cemeteries.

A widespread practice and its organisation

Very recent archaeological research comparing territories over a vast area of Europe, stretching from Transylvania to England, has highlighted the ubiquity of grave reopening and common patterns of this activity during the early medieval period (Klevnäs *et al.* 2021). This investigation has shown that the phenomenon appears to be concentrated between the 6th and the 7th centuries (thus, after people started to inhume the dead with grave goods), and that it commonly occurred within roughly a generation from the deposition, or even before. It was mainly directed towards the recovery of grave goods, but the motivations behind the violations were various and by no means only economic and for profit (see Lafferty 2014 on economic reasons). Research in the Netherlands has suggested that reopening should be seen as part of the complex relationship between communities and their dead (van Haperen 2010; 2017; 2018). Important previous studies have convincingly argued that grave robbers would have had some knowledge of the people buried and would have chosen where to dig on the basis of what they expected to find. Moreover, larger cemeteries were generally violated more often than smaller ones (Effros 2002 discussing previous scholarship). However, how such scavenging was organised remains difficult to reconstruct.

Recent research on early post-Roman Britain by Robin Fleming may show such an organisation, which very interestingly includes a top-down component. For instance, in the case of Cadbury Congresbury near Bristol, Fleming (2021, 84) points out that 'it is likely that the community undertook systematic scavenging campaigns in search of still-useable vessels at deserted sites in the neighborhood.' Cadbury Congresbury is a key site for post-Roman western Britain (Rahtz *et al.* 1992; Alcock *et al.* 1995; Davey 2005; Costen 2011). It is a prehistoric hillfort, which was later reoccupied in the 5th century – a common phenomenon in this area. It lay on the northern border of the Mendip Hills, which is one of the most important zones for mineral wealth on the island. A large hall and other structures indicate that Cadbury was an elite site from the 5th century until the end of the 6th/the beginning of the 7th century, when it became a monastic site. In the 6th century (and even slightly earlier), the recovery of Mediterranean ware, coins, glass and luxuries show that Cadbury was included into the Atlantic system that connected Ireland, western Britain, Spain, France and the Mediterranean. Importantly for the present study, even earlier (in the 5th century), the high-status community on this site had access to a certain amount of Roman pottery and glass – and, very interestingly, these items were used for dining. Concerning glass, at least 10 Roman glass bottles and beakers have been found; while not a huge number per se, it should be noted that only the 5% of the site has been excavated. Some of these items are dated to the 1st or 2nd centuries, while others date to the 4th century. The same is true for the pottery, among which there were vessels from the 1st and 2nd centuries. Fleming plausibly suggests that, while some of these materials may have been passed on as heirlooms, the oldest pots and glasses most likely came from burial grounds.

The area where Cadbury is located shows strong connections to high status and royal power throughout the early medieval period; other royal sites are documented and have been excavated nearby, and in general this presence may be connected to the resources of the area. Thus, if one follows Fleming's suggestion that the Roman items found in the 5th-century context were the output of 'systematic scavenging campaigns' that involved cemeteries, this 'organisation' may have been top-down to some degree, since the post-Roman power was involved. As these vessels and glassware were used by this high-status part of the population for dining, the value of these objects was indeed significant.

It is worth stressing that some level of organisation can thus be suggested in this area, which is notorious for the highly fragmentary state of the evidence of the earliest post-Roman centuries. This is indeed an interesting example of circuits that were likely active at the level of a specific community, with elites possibly involved in scavenging that was aimed at particular materials (as far as we can tell). However, the main issue with examples like this one is that they are episodic.

Scavenging of funerary remains could take on a variety of meanings, which are reflected in the written sources. In the religious context, for instance, this well-attested practice was sometimes praised as an act of devotion. Remaining in early medieval England, Bede (d. AD 735) narrated a relevant episode in relation to the reburying of the remains of Etheldreda, founder of the monastery of Ely in Cambridgeshire. Sexburg, the abbess of the community, decided that Etheldreda's body needed a new coffin. After searching for the right stone, people from the monastery 'came to a small ruined city,' and 'they discovered near the city walls a white marble sarcophagus of very beautiful workmanship with a close-fitting lid of similar stone.' Here, Bede seems to imply that the abbey could have made a new sarcophagus, as he stressed that the choice to go searching for one to reuse was dictated by the lack of access to stone the proper size in the area surrounding the monastery ('since the district of Ely was surrounded on all sides by sea and fens and had no large stones, they took a boat...') (transl. Sherley-Price 1995. See Liverani 2004; Bavuso *et al.* 2023).[6] Whatever the motivation, this story raises questions about the degrees of organisation of collection activities. This is important in order to understand how embedded reuse was in the local economy and the role of reuse in the economy generally.

The story of Etheldreda's coffin is one example, but many more can be identified. Yet, it is particularly difficult to formulate general models on the organisation of scavenging solely based on textual evidence. Again, and perhaps similar to Cadbury-Congresbury, the case of the sarcophagus of Etheldreda attests to a certain level of top-down organisation of the collection, albeit for an ad hoc event – this time within the confines of a monastery and its surrounding territory. The scavenging of cemeteries in more complex and coordinated forms is rarer and its organisation is more difficult to identify, especially because of the scarcity of written sources for this period. However, it seems likely that such activities indeed took place, at least in some areas of Europe, in addition to the organised scavenging of *villae*.

A top-down organisation, which involved central authorities and was thus based on much wider circuits than those visible from post-Roman England, is evidenced by sources for Ostrogothic Italy. There are incomparably more sources for Italy than England in this same period. For example, in a letter included in Cassiodorus' *Variae* and dated to *c.* AD 507–11, King Theoderic orders Duda, a *saio* (an official and royal envoy charged with the management of specifically limited tasks; see La Rocca and Tantillo 2017; Cohen 2022) to recover the gold and silver deposited at a burial site.

> King Theoderic to Duda, Saio.
>
> 1. It is the inclination of wisdom to recover for human use coins hidden in the earth and for the commerce of the living not to have contact with the dead, since things buried are lost to us and profits are lost to them in no measure. Indeed, the circulation of metal

> coins is the comfort of men. For if it lies neglected it is similar to veins of rich gold abandoned in the earth; it increases in value with use, when what is enclosed in greedy hands becomes properly disposed among the living. 2. And so, we have decided with a measured decree that you should convene with a public notary at that place where much wealth is accumulated near the surface, and if, as it is said, gold or silver may be uncovered by your searching, you may faithfully lay claim to it for public profit. Nevertheless, you should restrain the hand from the ashes of the dead, because we do not want to seek that wealth which is likely to be discovered by grave robbers (*per funesta scelera*). Let monuments cover the ashes, and columns and marble adorn the graves; let those who have abandoned the commerce of the living not retain wealth. 3. For gold is rightly removed from graves where an owner is not in possession. On the contrary, it is a kind of crime to uselessly abandon to the secrets of the dead that from whence the welfare of the living is able to sustain itself. It is not greed to rescue what no owner would grieve to lose [...] (Trans. Bjornlie 2019, modified).[7]

Although, as has recently been stressed, the letter had an episodic character (La Rocca and Tantillo 2017), it is worth noting that this was a royal command, and the collection appears to have been organised. In effect, the collection was carried out in the presence of the *saio* and other public officials (*sub publica testificatione*), and the recovered wealth was then claimed for the state (*compendio publico fideliter vindicabis*). Therefore, the state was involved in practices of collection, and the recovered materials mentioned in the letter – namely gold and silver – were indeed valuable to the state. Although this letter is a unique case within Cassiodorus' *Variae*, it can be connected to others within this corpus.

An example is another letter of Theoderic of about the same date written to the *comes* and *vir spectabilis* Anna. The letter reports that the king heard that a certain presbyter Laurentius 'has been searching among corpses of men for funerary riches by exhuming the remains, and he has inflicted harm upon those dead who ought to be shown respect from the living' (trans. Bjornlie 2019).[8] This is a completely different case of tomb violation, which may well have been an unorganised activity. In the letter, Theoderic ordered the *comes* to conduct an inquiry; however, Anna was to limit himself to the investigation only. Indeed, because of the 'dignity of the priesthood,' the crime would remain unpunished. As noted by Cristina La Rocca and Ignazio Tantillo (2017), this decision may have been framed within a general context of competition in this period in Italy between lay and religious institutions over the control of the funerary sphere.

As already mentioned, the violation of tombs was a practice condemned in the *edictum Theoderici*, where a person who destroyed a grave was sentenced to death (MGH LL 5, n. 110: *qui sepulchrum destruxerit, occidatur*; see Lafferty 2013). In the *Variae*, regulations surrounding cemeteries are also hinted at in a *formula* for the appointment of the *comes privatarum* (VI.8), an official who also appears to have been responsible for funerary sites. Like in the above-mentioned 4th-century laws, the stress in the *formula* is on building materials, such as marble and columns. As usual, violation of tombs was a sacrilegious act, and 'one who is revealed as having stolen something from the dead is deemed foreign to all sense of devotion' (trans. Bjornlie 2019; La Rocca and Tantillo 2017).[9]

Contamination is a recurring theme in the sources. In the letter for the *comes* Anna, we can read that 'it is claimed that such corrupt contact has not been withheld from hands dedicated to the sacred rites' (it is worth repeating that the culprit here was a presbyter). Contamination of the living was one of the two reasons why tomb violation was a crime in the edict of Constantius cited above. The second part of the edict by Julian could have had a similar implication: it prescribed that funerals should not be attended by crowds of people, as 'this practice polluted the eyes of men by its ill-omened aspect.'

Contamination also affected the goods recovered from burial grounds; hence the need to purify these items. This idea in effect explains the existence of complex ritual practices surrounding the reuse of ancient objects. This is well exemplified by the prayers included in the *Liber sacramentorum Gellonensis* (a text of the 8th century), which could be deployed in order to purify the vessels found in *loci antiqui*. These prayers may reflect a new concern about these objects being used in the liturgy from the 8th century (Effros 2001; Pazienza 2014).

> O God, you who have cleansed all the sins off your faithful people through the coming of your son our lord Jesus Christ, come hither, heeding our invocations with benevolence, and cleanse these vessels with the generosity of your grace, which you have returned to human use, extracted from the abyss of earth after the vastities of time for the indulgence of <your> mercy (*Liber sacramentorum Gellonensis* CCSL 159, 450).[10]

Returning to the letter addressed to the *saio* Duda, the characteristics of this crime – impiety and pollution – explain well the letter's convoluted explanatory discourse, which amounts to a justification or excuse for the king's promotion of scavenging. The reflections on the right to acquire goods neglected in the earth that had no owner echo yet another discourse about the morality of accumulated wealth that is found in a further letter included in the *Variae* (IX.3). In this letter, King Athalaric ordered his *comes patrimonii* to exploit the mines on the royal property in Bruttium. 'For why should that which might be an honest profit be allowed to lie without use? If indeed it is wrong to seek gold through war, through treachery on the sea, through scandal of falsehood, let it be found justly in nature itself. Gains by which no person grieves are honest, and what is taken from nobody's property is properly acquired' (trans. Bjornlie 2019; on this rhetoric, see La Rocca and Tantillo 2017, who, as mentioned above, interpret the letters in the context of competition between kings and the church).[11] Here, the reflection on the right to take for oneself properties without an owner is the same thinking expressed in the letter to Duda. Thus, the nature of the activity dictates a reflection on the act that the king is ordering, and – it is important to highlight – the letter to Duda takes pains to draw a boundary between this act and grave-robbing. The difference is stressed by the king's order that the dead are left untouched. Top-down tomb-robbing could be organised by the king and his agents, but they had to be protected from contamination and the risk of impiety.

Conclusions

The literary sources highlight important reflections, thoughts and, in some cases, justifications surrounding a practice that was seen as ambiguous at the very least. Since this contribution has focused on recycling and reuse, the complex link between scavenging and relics has been left aside. However, the *Liber sacramentorum* and the reuse of sarcophagi in connection with acts of devotion already allow us to see these practices in relation to cult activities.

While we have considerable evidence for the scavenging of cemeteries in the early medieval period, the economic and social organisation of this activity remains elusive. Yet, the literary sources have enabled us to explore a more general understanding of the involvement of power in this practice and to appreciate that beyond *villae* and other monumental sites, cemeteries were scavenged with varying degrees of organisation in each case. In the Ostrogothic kingdom, the letters to Duda and to Anna present the co-existence of organised collection and disorganised practices, and it is only the disorganised scavenging which should be punished. The organised practice is indeed a top-down one, involving central authorities and public officials, even if the nature of the activity dictates the need for some justifications. Early post-Roman England, on the other hand, can hardly compare to Italy in the same period, and the possibility that a similar level of organisation existed there is highly unlikely. Nevertheless, asking questions about the organisation of scavenging in burial grounds remains important even in this case. It is significant that elites may also have been involved in this practice in England.

The reuse of ancient materials and objects has largely been studied assuming a strict correlation with the economy. Alternatively, it has been studied with regard to the symbolic or amuletic roles of some of these items (*e.g.*, Meaney 1981; Effros 2003). As elusive (or indeed invisible) as it may be, investigating the entanglement of bottom-up and top-down organisation with regard to these activities – and how actors negotiated religious conceptions and ideas about contamination – may open new perspectives on the role of recycling and reuse in early medieval communities.

Notes

1 Οὔποτ' ἐγὼ φθιμένοισιν ἐπέχραον, οὐδ' ἀπὸ τύμβων / ἔργον ἔγειρα, δίκην ὄμνυμι καὶ φθιμένους· / τοὔνεκα μηδ' ἐπ' ἐμοῖσι φέρειν λάεσσι σίδηρον· / εἰ δὲ φέροις, τὴν σὴν ἐς κεφαλὴν πεσέτω. / Μαρτινιανὸς ἐγὼ τάδε λίσσομαι· εἴ τις ἐμεῖο / κύδεός ἐστι χάρις, τύμβος ἀεὶ μενέτω.

2 Χάζεο, χάζεο τῆλε· κακὸν τὸν ἄεθλον ἐγείρεις, / λᾶας ἀνοχλίζων καὶ τάφον ἡμέτερον· / χάζεο· Μαρτινιανὸς ἐγώ, καὶ ζῶσιν ὄνειαρ / καὶ νέκυς οὐκ ὀλίγον ἐνθάδε κάρτος ἔχω.

3 *CTh* IX.17.1: *Imp. Constantius a. ad Titianum praefectum urbi. Si quis in demoliendis sepulchris fuerit adprehensus, si id sine domini conscientia faciat, metallo adiudicetur; si vero domini auctoritate vel iussione urgetur, relegatione plectatur. et si forte detractum aliquid de sepulchris ad domum eius villamque pervectum post hanc legem repperietur, villa sive domus aut aedificium quodcumque erit fisci viribus vindicetur. dat. vii kal. iul. mediolano acindyno et proculo conss. (340 iun. 25).*

4 *CTh* IX.17.4: *Idem a. ad populum. qui aedificia manium violant, domus ut ita dixerim defunctorum, geminum videntur facinus perpetrare, nam et sepultos spoliant destruendo et vivos polluunt fabricando. Si quis igitur de sepulchro abstulerit saxa vel marmora vel columnas aliamve quamcumque materiam fabricae gratia sive id fecerit venditurus, decem pondo auri cogatur inferre fisco* [...] *(357 [356] iun. 13).*

5 *CTh* IX.17.5: *Imp. Iulianus a. ad populum. pergit audacia ad busta diem functorum et aggeres consecratos, cum et lapidem hinc movere et terram sollicitare et cespitem vellere proximum sacrilegio maiores semper habuerint. sed et ornamenta quidam tricliniis aut porticibus auferunt de sepulchris. quibus primis consulentes, ne in piaculum incidant contaminata religione bustorum, hoc fieri prohibemus poena manium vindice cohibentes. (363 febr. 12).*

6 Bede, *Historia ecclesiastica gentis anglorum* IV.19: *iussitque quosdam e fratribus quaerere lapidem, de quo locellum in hoc facere possent; qui ascensa naui, ipsa enim regio Elge undique est aquis ac paludibus circumdata, neque lapides maiores habet, uenerunt ad ciuitatulam quandam desolatam, non procul inde sitam, quae lingua Anglorum Grantacaestir uocatur; et mox inuenerunt iuxta muros ciuitatis locellum de marmore albo pulcherrime factum, operculo quoque similis lapidis aptissime tectum.* [...] *Mirum uero in modum ita aptum corpori uirginis sarcofagum inuentum est, ac si ei specialiter praeparatum fuisset; et locus quoque capitis seorsum fabrefactus ad mensuram capitis illius aptissime figuratus apparuit.*

7 MGH AA 12, Cassiodorus, *Variae*, IV.34 (c. 507–11): *Duda saioni Theodericus rex. Prudentiae mos est in humanos usus terries abdita talenta revocare commerciumque viventium non dicere mortuorum, quia et nobis infossa pereunt et illis in nullam partem profutura linquuntur. Metallorum quippe ambitus solacia sunt hominum. Nam divitis auri vena similis est reliquae terrae, si iaceat: usu crescit ad pretium, quando et apud vivos sepulta sunt, quae tenacium manibus includuntur. Atque ideo moderata iussione decernimus, ut ad illum locum, in quo latere plurima suggeruntur, sub publica testificatione convenias: et si aurum, ut dicitur, vel argentum fuerit tua indagatione detectum, compendio publico fideliter vindicabis: ita tamen ut abstineatis manus a cineribus mortuorum, quia nolumus lucra quaeri, quae per funesta possunt scelera reperiri. Aedificia tegant cineres, columnae vel marmora ornent sepulcra: talenta non teneant, qui vivendi commercia reliquerunt. Aurum enim sepulcris iuste detrahitur, ubi dominus non habetur: immo culpae genus est inutiliter abditis relinquere mortuorum, unde se vita potest sustentare viventium. Non est enim cupiditas eripere, quae nullus se dominus ingemiscat amisisse.* [...]

8 MGH AA 12, Cassiodorus, *Variae*, IV.18 (c. 507–11). [...] *dudum siquidem ad nos multorum suggestione pervenit Laurentium presbyterum effossis cineribus funestas divitias inter hominum cadavera perscrutatum concussionemque mortuis intulisse, quem oportet viventibus quieta praedicare. non abstinuisse perhibetur tam crudeli contagio piis dicatas consecrationibus manus: aurum exsecrabili quaesisse fertur affectu, quem suam decuisset egentibus dare substantiam vel sub aequitate collectam. Quod te diligenti examinatione praecipimus indagare, ut, si veritati dicta perspexeris convenire, hominis ambitum eo tantum fine concludas, ne possit supprimere quod eum non licuit invenire. scelus enim, quod nos pro sacerdotali honore relinquimus impunitum, maiore pondere credimus vindicandum.*

9 MGH AA 12, Cassiodorus, *Variae*, VI.8: [...] *Defunctorum quin etiam sacram quietem aequabilia iura tuae conscientiae commiserunt, ne quis vestita marmoribus sepulcra nudaret, ne quis columnarum decorem inreligiosa temeritate praesumeret, ne quis cineres alienos aut longinquitate temporis aut voraci flamma consumptos scelerata perscrutatione detegeret, ne corpus, quod semel reliquerat molestias mundanas, humanas iterum pateretur insidias. Nam etsi cadavera furta non sentiunt, ab omni pietate alienus esse dinoscitur, qui aliquid mortuis abrogasse monstratur. vide quae tibi commissa sunt: castitas viventium et securitas mortuorum.* [...]

10 *Deus qui aduentu[m] fili tui domini nostri iesu christi omnia tuis mundasti fidelibus, adesto propitius inuocationibus nostris, et haec uascula que* <*tuae*> *indulgentiae piaetatis post spatia temporum a uoragine terrae abstracte humanis usibus reddedisti, gratiae tuae largitate aemunda.*

11 MGH AA 12, Cassiodorus, *Variae*, IX.9 (c. 527): [...] *cur enim iaceat sine usu, quod honestum potest esse compendium? Aurum si quidem per bella quaerere nefas est, per maria periculum, per falsitates opprobrium, in sua vero natura iustitia. honesta sunt lucra per quae nemo laeditur et bene adquiritur quod a nullis adhuc dominis abrogatur.* [...]

Bibliography

Abbreviations

MGH = *Monumenta Germaniae Historica*
CTh = *Theodosiani libri XVI cum Constitutionibus Sirmondianis et Leges Novellae ad Theodosianum pertinentes*

Primary sources

Bede. *Ecclesiastical History of the English People: with Bede's Letter to Egbert and Cuthbert's Letter on the Death of Bede*. Translated by Leo Sherley-Price. London, Penguin, 1990.

Bede. *Historia ecclesiastica gentis anglorum*. Edited by Charles Plummer, *Venerabilis Baedae Opera Historica*. 2 vols. Oxford, Clarendon, 1896.

Bjornlie, M.S. (2019) *The Variae: The Complete Translation*, Oakland, University of California Press.

Cassiodorus. *Variae*. Edited by Theodor Mommsen, MGH, *Auctores Antiquissimi* 12. Berlin, Weidmann, 1894.

Edictum Theoderici. Edited by Friedrich Bluhme, MGH, *Leges* 5. Hannover, Hahn, 1889.

Lex Salica. Edited by Karl August Eckhardt, MGH, *Leges nationum Germanicarum* 4.1, *Pactus Legis Salicae*. Hannover, Hahn, 1962.

Liber Sacramentorum Gellonensis. Edited by Antoine Dumas, *Corpus Christianorum Series Latina* 159. Turnhout, Brepols, 1981.

Palatine Anthology. Edited by William R. Paton. Revised by Michael A. Tueller, *The Greek Anthology*. Cambridge, MA, Harvard University Press, 2014.

Pharr, C. (1952) *The Theodosian Code and Novels and the Sirmondian Constitutions. A Translation with Commentary, Glossary, and Bibliography*. New York, Greenwood Press.

Theodosiani libri XVI cum Constitutionibus Sirmondianis et Leges Novellae ad Theodosianum pertinentes. Edited by Theodor Mommsen, Paul M. Meyer. 3 vols. Berlin, Weidmann, 1905.

Secondary sources

Alcock, L. with Stevenson, S.J. and Musson, C.R. (1995) *Cadbury Castle, Somerset. The Early Medieval Archaeology*. Cardiff, University of Wales Press.

Barfod, G.H., Feveile, C. and Sindbæk, S.M. (2022) Splinters to splendours: from upcycled glass to Viking beads at Ribe, Denmark. *Archaeological and Anthropological Sciences* 14, 180.

Bavuso, I., Furlan, G., Intagliata, E.E. and Steding, J. (2023) Circular economy in Roman and early medieval Europe – methods of analysis for a future agenda. *Open Archaeology* 9(1), 20220301. https://doi.org/10.1515/opar-2022-0301.

Cohen, S. (2022) Gelasius and the Ostrogoths: jurisdiction and religious community in late fifth-century Italy. *Early Medieval Europe* 30.1, 20–44.

Costen, M. (2011) Early settlement around the Mendips: place-names and local history. In J. Lewis (ed.) *The Archaeology of Mendip: 500,000 Years of Continuity and Change*, 257–274. King's Lynn, Heritage Marketing.

Davey, J.E. (2005) *The Roman to Medieval Transition in the Region of South Cadbury Castle, Somerset*. Oxford, British Archaeological Reports.

Duckworth, C.N. and Wilson, A. (eds) (2020) *Recycling and Reuse in the Roman Economy*. Oxford, Oxford University Press.

Effros, B. (2001) Monuments and memory: repossessing ancient remains in early medieval Gaul. In M. de Jong, F. Theuws and C. van Rhijn (eds) *Topographies of Power in the Early Middle Ages*, 93–118. Leiden, Brill.

Effros, B. (2002) *Caring for Body and Soul. Burial and the Afterlife in the Merovingian World*. University Park, Pa., Pennsylvania State University Press.

Effros, B. (2003) *Merovingian Mortuary Archaeology and the Making of the Early Middle Ages*. Berkeley, University of California Press.

Fleming, R. (2021) *The Material Fall of Roman Britain, 300-525 CE*. Philadelphia, University of Pennsylvania Press.

Floridi, L. (2013) The epigrams of Gregory of Nazianzus against tomb desecrators and their epigraphic background. *Mnemosyne* 66, 55–81.
Goldhill, S. and Greensmith, E. (2020) Gregory of Nazianzus in the *Palatine Anthology*: the poetics of Christian death. *The Cambridge Classical Journal* 66, 29–69.
Hodges, R., Leppard, S. and Mitchell, J. (eds) (2011) *San Vincenzo Maggiore and its Workshops*. London, British School at Rome.
Klevnäs, A., Aspöck, E., Noterman, A.A., van Haperen, M.C. and Zintl, S. (2021) Reopening graves in the early Middle Ages: from local practice to European phenomenon. *Antiquity* 95.382, 1005–1026.
La Rocca, C. and Tantillo, I. (2017) Corredi, corpi e reliquie nelle *Variae* di Cassiodoro. La competizione tra re e vescovi per le risorse del sottosuolo. In V. Loré, G. Bührer-Thierry and R. Le Jan (eds) *Acquérir, prélever, contrôler: les ressources en compétition* (400–1100), 21–42. Turnhout, Brepols.
Lafferty, S. (2013) *Law and Society in the Age of Theoderic the Great. A Study of the* Edictum Theoderici. Cambridge, Cambridge University Press.
Lafferty, S. (2014) *Ad sanctitatem mortuorum*: tomb raiders, body snatchers and relic hunters in late antiquity. *Early Medieval Europe* 22.3, 249–279.
Liverani, P. (2004) Reimpiego senza ideologia. La lettura antica degli spolia, dall'Arco di Costantino all'età di Teodorico. *Mitteilungen des Deutschen Archäologischen Instituts - Römische Abteilung* 111, 383–434.
Marazzi, F. and D'Angelo, F. (2009) Il ciclo della produzione vetraria a San Vincenzo al Volturno nel IX secolo: riflessioni da una rilettura dei dati archeologici. In J.-P. Brun (ed.) *Artisanats antiques d'Italie et de Gaule: Mélanges offerts à Maria Francesca Buonaiuto*, 211–223. Napoli, Centre Jean Bérard.
Meaney, A.L. (1981) *Anglo-Saxon Amulets and Curing Stones*. Oxford, British Archaeological Reports.
Moroni, M.G. (2013) Gli epigrammi di Gregorio Nazianzeno contro i violatori di tombe II. Aspetti esegetici e letterari. In V. Zimmerl-Panagl (ed.) *Dulce Melos II. Arten des 5. Internationalen Symposiums: Lateinische und griechische Dichtung in Spätantike, Mittelalter und Neuzeit, Wien, 25.-27. November 2010*, 47–66. Pisa, Edizioni ETS.
Munro, B. (2020) The organized recycling of Roman villa sites. In C.N. Duckworth and A. Wilson (eds) *Recycling and Reuse in the Roman Economy*, 383–401. Oxford, Oxford University Press.
Pazienza, A. (2014) Identity, funerary practices and memory in Lombard Tuscia (6th to 8th Centuries). *Zeitschrift für Archäologie des Mittelalters* 42, 1–31.
Rahtz, P., Woodward, A., Burrow, I., Everton, A., Watts, L., Leach, P., Hirst, S., Fowler, P. and Gardner, K. (1992) *Cadbury Congresbury 1968–73: A Late/Post-Roman Hilltop Settlement in Somerset*. Oxford, British Archaeological Reports.
Rebillard, É. (2003) *Religion et sépulture: L'Église, les vivants et les morts dans l'Antiquité tardive*. Paris, École des hautes études en sciences sociales.
Schibille, N. and Freestone, I.C. (2013) Composition, production and procurement of glass at San Vincenzo al Volturno: an early medieval monastic complex in southern Italy. *PLOS ONE* 8.10.
Ubl, K. (2017) *Sinnstiftungen eines Rechtsbuchs: die Lex Salica im Frankenreich*. Ostfildern, Jan Thorbecke Verlag.
Van Haperen, M.C. (2010) Rest in pieces: an interpretive model of early medieval 'grave robbery'. *Medieval and Modern Matters. Archaeology and Material Culture in the Low Countries* 1, 1–36.
Van Haperen, M.C. (2017) *In Touch with the Dead. Early Medieval Grave Reopenings in the Low Countries*. Unpublished thesis, Leiden University.
Van Haperen, M.C. (2018) Exchanges with the dead: the economic aspects of reopening graves. In M. Kars, R. Van Oosten, M. Roxburgh and A. Verhoeven (eds) *Rural Riches and Royal Rags? Studies on Medieval and Modern Archaeology, Presented to Frans Theuws*, 110–114. Zwolle, SPA-Uitgevers/Amsterdam, Dutch Society for Medieval Archaeology.
Van Wersch, L., Verslype, L., Strivay, D. and Theuws, F. (eds) (2019) *Early Medieval Tesserae in Northwestern Europe*. Bonn, Habelt-Verlag.
Wilkin, A. (2019) Le travail et commerce du verre au haut Moyen Âge, un révélateur pour l'histoire économique? In L. Van Wersch, L. Verslype, D. Strivay and F. Theuws (eds) *Early Medieval Tesserae in Northwestern Europe*, 14–25. Bonn, Habelt-Verlag.

Chapter 3

Exploring reuse in a prestige environment: the palace city of Samarra

Rhiannon Garth Jones

It is increasingly clear that the cities of the past practiced economic circularity broadly and reuse practices specifically throughout their lifetimes. Recent research like that produced by the Circular Economy and Urban Sustainability in Antiquity project and at the Centre for Urban Network Evolutions has made significant developments both in individual classes of materials, such as coinage (Birch *et al.* 2019), glass (Boschetti *et al.* 2021; Barfod *et al.* 2022), metals (Orfanou *et al.* 2020) and stone (Intagliata 2021), as well as in broader theoretical approaches (Andrade and Raja 2022; Bavuso *et al.* 2023). In connection with a more nuanced theoretical approach to circular practices in the modern day, these developments increasingly suggest that circular practices should be seen as *part of* earlier economic systems even where those systems were motivated by different concerns than present day efforts, and/or where linear models of consumption also existed and/or where limited resources were not a factor (Furlan and Andreatta 2023). Although research on economic circularity has so far primarily focused on Roman and early medieval economies (Peña 2007; Duckworth and Wilson 2020a), some important work is now emerging in early Islamic archaeology (Birch *et al.* 2019; Intagliata 2021; Barfod *et al.* 2022). The recent *Journal of Islamic Archaeology Special Issue* on 'Refuse and Reuse in Islamic Archaeology' is a welcome development, but as José C. Carvajal López (2022, 2) noted in his editorial, it is an incredibly broad field and we have barely begun to scratch the surface of applying recent theoretical advances. Moreover, when reuse has been studied in the context of Islamic Archaeology, it has tended to be as part of a high-definition analysis of a single class of material in a single location (Barfod *et al.* 2018; Orfanou *et al.* 2020), and most often with a focus on building material and sculpture, such as in Umayyad Palmyra (Intagliata 2021) or 12th-century northern Syria (Gonnella 2010). A more holistic approach to economic circularity in an Islamic city has not yet been taken, nor has the Abbasid period received much attention.

Economic circularity is characterised by the reduction of waste and is therefore often understood as a pragmatic approach to limited resources (Winans *et al.* 2017, 826; Furlan 2023, 343). Research is often focused on places and periods associated with decline and/or resilience. Moreover, when reuse in particular by elites is studied, it is more often understood through the lens of symbolism and associations with or demonstrations of power. Even where the financial element is considered – at the level of government or public function as opposed to the private level (Intagliata 2021, 435) – it is weighed up against the potential meanings of the material used rather than understood in terms of economic circularity (Gonnella 2010; Johnson 2017; Brown 2022). A consideration of economic circularity in elite environments adds to the complexity of our understanding and shifts our understanding of reuse in pre-modern economies further towards a standard practice, rather than seeing it as a specific response to straitened circumstances.

The Abbasid imperial capital of Samarra offers an ideal case study to begin to address these two gaps in current research on circular economy – elite behaviour in the Abbasid period and a more holistic view of a range of practices within one city. This chapter provides an overview of circular economy practices in the city during the 56-year period it served as the capital of the Abbasid Caliphate at the peak of its power – often considered part of the Islamic 'Golden Age'. I work from the starting point that the *concepts* of consumerism and linear practices are opposed to the *concept* of circular practices (although the two approaches can be practiced concurrently). The former is 'indissolubly tied up with the idea of purchasing something new' (Furlan forthcoming). The latter is characterised by the reduction of waste and the avoidance of using new resources when existing resources are available and includes practices like sharing, reuse, repair and recycling. I employ the methodology that has proved fruitful in a number of recent studies, carefully analysing texts to inform and rethink the gaps in the archaeological record and reconsidering the available data (Duckworth and Wilson 2020a; Bavuso *et al.* 2023). This overview is not comprehensive, but it provides a foundation to frame future discussions and identifies questions for further research.

Imperial Samarra

Samarra is located on the east bank of the Tigris, 125 km north of Baghdad. It was chosen by caliph al-Mutasim (AD 796–842) as the site for his new imperial capital in AD 836, having been previously used by the Abbasids and Sasanians before them as a hunting ground. In AD 892, the government returned to Baghdad, and Samarra shrank in size and significance, although a mint remained and the tombs of the eleventh and twelfth Shia imams made it a site of pilgrimage from the 940s to the present day (I use the Gregorian calendar here to better support continuity throughout the edited volume; the time range here corresponds to 221–278 AH). Samarra is a city often associated with decadence and then rapid decline (Muqaddasi, *trans.* in Collins

1994, 111). Caliph after caliph built new extravagant palaces before either they or their heirs moved onto a new project (Northedge 2001, 46). Luxury decorative items were brought from all over the empire to demonstrate the caliph's power and prestige, although very little material trace of them remains (Saba 2022). It was an economic superpower that became a poetic symbol of loss (Yaqut cited in Meisami 2001, 69).

Although the focus here is on the imperial period, it is important to note that the Abbasid capital was founded on a site that had been settled as early as the Chalcolithic period (Northedge 2005, 43). It was a significant Sasanian site before the Arab Conquest of Mesopotamia, and a settlement remained after the caliphs relocated back to Baghdad – and remains to this day. The site has been thoroughly scavenged over time – so it has been a significant location for reuse economies practiced elsewhere, an often-overlooked element of the site.

Samarra is a huge archaeological site with ruins including hunting grounds and different urban centres spread across more than 57 km^2. The earliest excavations were carried out by archaeologist Ernst Herzfeld in 1911 and 1912–13; at the end of this, Herzfeld had recorded 1161 finds including ceramics, glass, wall painting, stucco and other media (Kadoi 2014, 28). Considerable work has been done to supplement Herzfeld's research since, starting with the many excavations at the site that have been conducted by the Iraq Directorate-General of Antiquities (Al-Jannabi 1982). Since the 1980s, the main source of new archaeological information on the site has been the surface surveys produced by Alastair Northedge (2005). By combining photogrammetric techniques from aerial photography with work on the ground, supplemented with textual sources, Northedge (2001) has been able to construct a detailed city plan.

Imperial Samarra does not fit the usual paradigm to explore reuse practices. Estimations of the cost of palace building at the city vary wildly; although, Northedge (2005, 258) has suggested it is likely that, for instance, 'up to 256/870 at least, a large proportion of the state budget was being spent in Samarra by the court and the army' on building projects for the new city. However, there is no real suggestion that the caliphs were troubled by financial considerations at the time. As the home of the caliphs and their armies, as well as a site of regular monumental building, the city became an immediate centre of economic activity and a draw for craftspeople from as far away as China (Kennedy 2005, 147–148). It could be argued that it is a pre-eminent example of a pre-modern consumptive economy – with each new building project consuming new resources and driving economic processes, only to move onto another. However, the Abbasid caliphs in this period also practiced various forms of reuse and other practices that reduced waste.

Methodology

At imperial Samarra, the problem that 'recycling economies, when efficient, are by their nature invisible in archaeological terms' (Needham *et al.* 1989, 384) is compounded by a number of factors, including the different types of removal of materials that

I will discuss in more detail later. However, the written sources provide clues when re-read from the particular perspective of circularity (Bavuso *et al.* 2023). Although these sources do not address circularity directly, likely for the very reason that such practices were assumed to be routine rather than noteworthy, they can, when used carefully, complement the material remains we do have and shed light on a number of types of circular practice.

There are a number of types of circularity that we might infer being practiced by the citizens of imperial Samarra, just as elsewhere: not least, the significant presence of soldiers in the city would suggest repair practices (Wild 2020, 67). There are also, of course, the various types of reuse and recycling presumably practiced by those who have scavenged the site over the years. This chapter, however, focuses on top-down circular practices: this is where we see the caliphs acting as agents. The bulk of archaeological excavation at imperial Samarra has focused on the urban planning and the caliphal buildings, in part because the site presents an opportunity to explore these two topics to a degree unmatched elsewhere (Millwright 2017, 177). The focus of the texts is also generally on the lives of the elite. It is, of course, always possible to read against the grain and reveal the lifestyles of non-elites in the texts; however, that is not the aim of this article. I use the texts to build on recent research that questions the assumptions that circular practices are only seen in moments of economic decline (Bavuso *et al.* 2023, 8).

Definitions

Increasingly, the terms associated with economic circularity are developing into a genre of literature on their own (Winans *et al.* 2017; Bavuso *et al.* 2023). A number of the examples I highlight fall into straightforward categories of, for example, recycling and reuse. Some are more nuanced, however, partly because this is a new area of research. The focus, however, is on the reduction of new resource usage and waste. More precisely, I employ: recycling (for example, building materials, glass, coins, weaponry); conservatory processes (for example, stewardship/curation and maintenance/repair (textiles)); and reuse, within which are

- Relocation (including, for example, of purpose-built items such as teak and marble panels intended for the same use in new venues, by the same caliph).
- Lateral cycling (for example, passed on to a new owner for the same use: see above).
- Secondary use (for example, Cup of Pharoah fountain, gates, etc.; booty, although this might also be lateral cycling, depending on the object in question).

The category of 'relocation' is not commonly used in studies of economic circularity and thus bears some explanation here. It is similar to 'sharing', but I use it here to cover the many movable items that decorated the imperial palaces and were intended to be used again and again, by the same caliph and his heirs, but in different palaces (and, significantly, within the same city rather than in the peripatetic kingship model practiced by their Umayyad predecessors or many medieval kings in Europe). This practice does not have a clear corollary. The items do not otherwise neatly fall into the more common categories, but they require inclusion, not least to contextualise practices that reduce waste in this seemingly consumeristic imperial capital.

The architectural décor at Samarra is mostly associated with cheap stucco panels, but it is unlikely that they would be the main element viewed by most audiences (Saba 2015; Millwright 2017, 179). Instead, the audience spaces would have been transformed by marble panels, teak panels, textiles, weaponry and other ornaments that moved with the court. Per Marcus Millwright (2017, 180), 'it was standard practice in royal courts to remove marble veneer and other valuable architectural fittings along with all the furnishings.' Only the stucco remains because the other decorative items were always intended to be relocated to the next palace for the next occasion or, indeed, to the next city. The assumption of a type of reuse was built into the prestige ornamentation that was used to decorate caliphal palaces (a type of reuse, moreover, which leaves very little material traces at the site). With that assumed reuse and relocation, we can surely also presume some amount of maintenance and repair, topics to which I will return.

Building materials

First, perhaps the most familiar group of materials, those larger items that have traditionally been considered as *spolia* (a problematic term but one pervasive in modern literature; see Dumser 2018), and what they might tell us about broader resource management by the caliphs at Samarra. Abbasid interest in antiquity is well-documented, as an articulation of power and continuity with past rulers (Hoffman 2008), as a source of *'aja'ib* or wonders (Johnson 2017), as places of healing (Gonnella 2010) and as sources of didactic or moral power (Mulder 2021, 15). Some work has also been done on the use of *spolia* as building materials in the early Islamic period in both a private and public capacity (Intagliata 2021), and at imperial Samarra (Brown 2022). There is good reason to believe, in the case of imperial Samarra, that the examples I highlight here were chosen for symbolic purposes.

Gates

Al-Mutasim, who moved the imperial capital to Samarra in AD 836, took a set of monumental iron gates from the city of Ammuriye, Turkey, and installed them in a palace at Samarra in AD 838 (Meinecke 1995, 412). This was common practice. 'In the medieval Near East, the transfer of gates from one city to another may have had a certain practical significance, but it was also a symbolic act expressive of authority' (Lassner 1980, 181). In fact, one of the gates was reportedly moved to al-Raqqa after Samarra was abandoned and thereafter to Aleppo (Meinecke 1995, 412). Matthew Saba (2022) has argued that Abbasid gates were stages for public performances of power, and medieval Arabic sources highlight the perceived significance of gates' associations in Baghdad (al-Tabari *trans.* Kennedy 1990, 5; al-Baghdādī *trans.* in Lassner 1970, 54). The symbolism is clearly significant. There are some possible practicalities, however: the craftsmanship and labour required to produce monumental gates was substantial, and the logistics of transporting them across distances was perhaps less of a drain on resources than building new ones (Heidemann 2006). An emerging theme of this survey is the likelihood of both symbolic and circular reasons for reuse.

Roman stone basins for water features

At the centre of the Great Mosque in Samarra, built by al-Mutawwakil in AD 848–852, was a famous water fountain (Yaqubi, trans. Fishbein 2018, 98–99) that incorporated a large stone basin made from non-local stone. It was called *kasat firun* or Cup of Pharoah. It reportedly had a circumference of *c.* 11.5 m and a diameter of 3.66 m (Brown 2022, 2). In 1922, Herzfeld uncovered a stone basin (Fig. 3.1) that matched this description in the Great Mosque's courtyard area, surrounded by fragments of marble columns, painted stucco and mosaic, which suggested a richly decorated area to match the significance implied by the literary sources. A similar, slightly smaller stone basin was later discovered in the courtyard of the Dar al-Khilafa, a large palace complex used by

Fig. 3.1 The stone basin discovered in the Dar al-Khilafa at Samarra by Ernst Herzfeld, FSA A.6 04.GN.3190. (Freer Gallery of Art and Arthur M. Sackler Gallery Archives, gift of Ernst Herzfeld, 1946).

the caliphs (Dahmani 2021). Moreover, other *kasat firun* are mentioned in the literary sources, including one that was apparently transported to Baghdad in AD 1256 (Brown 2022, 4). Peter J Brown has argued that 'Cup of Pharaoh' was a name in wider usage for these large stone basins and cites similar reuse of stone architectural elements as part of water features in, for example, the mosque of Ibn Tulun in Egypt (built AD 879) as well as in the new Umayyad dynasty at Al-Andulus in the 10th century. Both Fatma Dahmani and Brown have demonstrated that the basins found at Samarra were more likely to be Roman labra, stone water basins (Dahmani 2021; Brown 2022, 6). This has led to some confusion in the scholarship about the name. However, firʿawn in this period might have had a looser meaning of 'oppressor' (Wendell 1971, 116), presumably based on the Qur'anic stories of the prophet Musa, and thus the name might mean 'Roman' in this context. Brown (2022, 8) suggests both a symbolic and practical reason for the popularity of these stone basins: a legitimation for new dynasties of their claims to power as well as the difficulties of producing large basins from attractive stone. This explanation would hold whether the association was Egyptian or Roman.

In both examples discussed here, symbolism appears to be a factor in the selection of materials. However, resource management appears to be a plausible factor as well, albeit not necessarily the more significant one. I move now to other building materials that might fit more closely into an economic paradigm of reuse.

Former canal structures

The location of Samarra on the eastern bank of the River Tigris and its raised elevation meant that a water supply could not be drawn to it by gravity since its location was higher than the river and even higher than the *Qatul Kisrawi* canal adjoining to it from the east. Irrigation was therefore a necessity. It was also, in Mesopotamia, a traditional demonstration of power to be able to control water through irrigation, a particularly interesting factor to consider within the framework of circular economy. Al-Mutasim focused his main irrigation work on reconstructing and expanding the *Nahr Ishaqi* canal, previously built and used by both the Parthians and the Sasanians at the site. All but one of the brick-built bridges at Samarra (*qantara*), constructed by al-Mutasim and his successors, were made from the fired bricks of Sasanian aqueducts, reused and repaired or rebuilt to allow humans to cross the canals (Northedge 2005, 148).

Unseen building materials

The vast scale of building projects at Samarra was made possible not only by the financial resources of the caliphs but also by the huge amount of empty space combined with cheap (and easily degraded) building materials available. For the most part, mud-brick was used to build the many caliphal palaces, sometimes the more expensive fired type (perhaps for the more important sections, certainly for the Great Mosque), often unfired, and then decorated – first cheaply with stucco and paint, and then with more lavish items, which I will discuss below. In Baghdad, we are told by al-Tabari that al-Mansur, the founder of Baghdad and second

Abbasid caliph, had considered reusing materials from nearby Sasanian Ctesiphon in the city as a demonstration of his power but changed his mind after exploring the financial cost of such a decision. He chose the cheaper option of local materials instead (al-Tabari *trans.* Kennedy 1990, 4–5). This is a particularly interesting anecdote for a number of reasons, and it is not the only time that al-Mansur is shown to be extremely concerned with fiscal prudence in the building of Baghdad. We can infer, then, that reuse in the early Abbasid capital cities was complex and not always driven by symbolism and expressions of power – resource management could also be a factor.

Millwright has considered the various building materials used or likely to have been used at imperial Samarra and speculated about their possible origin. For instance, local palm wood might have been used in less visible places (although Yaqubi claims more was required than was available locally), and it is likely that sand and other additives needed for stucco could be taken from the banks of the Tigris (Millwright 2001, 86). As for the fired bricks, he highlights two possibilities: A) that they were taken from the various pre-Islamic (Sasanian) structures nearby, transported to Samarra and reused in new structures; or B) they were manufactured from scratch, requiring significantly more fuel – at least one likely source of which would be agricultural waste (Millwright 2001, 85). In the case of A, the need for fuel in making new bricks would be reduced, but the process of extraction could be difficult and would require more manpower. In the case of B, both the length of time and the volume of fuel required would be substantial, although the workforce required would be slightly reduced.

I would suggest a third option: that both practices could have been used, with B supplementing A. Given the availability of significant amounts of manpower to the caliphs, it seems more likely that reuse of material was the primary practice and that the more expensive new burnt bricks would have provided additional materials (Millwright 2001). The literary sources give no indication, and the state of the remains at Samarra make it very hard to calculate (Millwright bases his calculations on other sites combined with 18th-century records from the region and modern firing kilns). However, it seems *some* circularity must have been practiced in creating the fired bricks that supplemented the cheaper, easier, lower quality unfired bricks, whether reuse of the existing bricks, recycling of waste for fuel or a combination.

Decorative materials

The first layer of decoration in the palaces was cheap stucco panels, as I have discussed above. Layered on top of that basic structure, however, was a dazzling array of decorative items to project power and provoke wonder in the eyes of the audiences (Saba 2015). A description of a visit to Baghdad from an ambassador from Constantinople in AD 917, which is considered representative of the palaces at Samarra, describes thousands of textiles, armour, many different types of weapons and other portable items of silver and gold brought out from the caliphal treasury to decorate the space (Millwright 2017, 179). We also know that teak (Fig. 3.2) and marble panels

Fig. 3.2 Carved and painted wooden panel found in the Audience Hall Complex (Museum inventory number: Sam 925); 61 cm (height) × 924 cm (width) × 4.8 cm (max. depth). Such panels were likely moved from palace to palace with the caliphs. (Staatliche Museen zu Berlin, Museum für Islamische Kunst / photographer: Christian Krug).

would have been transferred from palace to palace to supplement the visual effect (Saba 2015). The written sources give us an insight into the 'profoundly adaptable quality of the architectural spaces in early Islamic Palaces' (Millwright 2017, 179) but read with a focus on circular practices, they also provide some insight into behaviour.

Luxury textiles

Early Islamic textiles are poorly preserved, rarely studied at all and not yet studied for signs of maintenance and repair (Dospěl Williams 2019, 411). We know from a number of literary sources that luxury textiles were extremely important in decorating spaces, perhaps the most important element (Mackie 2015, 29), as they were able to conceal and reveal other aspects of the interior decoration (Saba 2015) and included cushions and carpets (Awwad 2008). They were also frequently multipurpose, interchangeable not just in terms of object to object but object to person (Mackie 2015, 29). Their value was such that they might also have been used instead of money, including to pay government taxes (Mackie 2015, 41), and they were part of the elite culture of luxury goods that allowed rulers in antiquity and the medieval period to demonstrate their wealth and construct royal identities. In the Abbasid court, luxury textiles included carpets, wall hangings, curtains and personal robes. They were an integral part of the court décor – arguably, the most important aspect. In fact, the early Abbasids were 'known for their textiles' (Mackie 2015, 70). Chinese weavers worked in the city during the 8th century, and we know of weavers from Iran and Egypt under al-Mutasim (Kennedy 2005, 147–148). Textiles found at Samarra and dated by one fragment of *tiraz* to the reign of al-Muntasir (AD 861–862) have a weave that is 'without parallel,' and they are 'characterized by resplendence and simplicity' (Mackie 2015, 70). They included threads made from gold foil wrapped around a silk. These were prestige items that were reused again and again by different caliphs in different settings (Pl. 1). For example, at the feast for the circumcision of al-Muʻtazz, held by al-Mutawakkil and described by al-Shabushti, the only carpet large enough for the Audience Hall was one plundered from the Umayyads, a 'carpet with thread like silk, embroidered, woven with gold and lined with fringe' (Shabushti *trans.* Awwad 2008, 150–151). For instance, al-Shabushti reports that:

> When (Mutawakkil) decided to hold a feast for the circumcision of Muʻtazz, he placed al-Fath ibn Khaqan in charge of the preparations and searching the treasuries of furnishings for a carpet that would fit the length and breadth of the iwan (a significant monumental structure), whose area was 100 cubits long and 50 cubits wide. There was none that fit except for one that had been plundered from the Umayyads, found among the possessions of Hisham ibn 'Abd al-Malik. It was a carpet with thread like silk, embroidered, woven with gold and lined with fringe...

If the report is accurate, this carpet would have been at least 100 years old and traced back through 14 reigning caliphs. Even if it is an exaggeration, it is clear that some valuable carpets were passed on for generations. We know, for example, that this behaviour is true elsewhere; marriage documents found in the Cairo Geniza document give extensive lists of textiles and furnishings in women's possessions, including their monetary values (Goitein 2020). It seems impossible that no conservatory practices would have been involved (very little research has been done on luxury Islamic textiles; *cf.* the discussion on Roman textiles in Wild 2020). It is noteworthy that non-durable Islamic textiles are known to have been criticised (see discussion in Mackie 2015, 31). Lastly, we know that precious gowns could be regifted from a caliph to a favoured courtier to show their status, a form of lateral cycling (for more on the redistributive society as exemplified by textiles, see Mackie 2015).

Glass tesserae

Nearly 400 glass finds have been catalogued at Samarra, including include mosaic tesserae, drawn tesserae and hollow inlays of colourless, transparent glass in a variety of shapes: diamond, triangle, circular and oval. Traces of stucco on some pieces suggest they may have been embedded into the stucco panels to refract light, a technique known elsewhere. The origins of these finds have been discussed as part of a high-resolution analysis performed by Nadine Schibille *et al.* This analysis demonstrated three different sources of supply: reused tesserae of Levantine and Egyptian origins, imports of contemporary coloured glass from the Levant and/or Egypt, and local glass production (Schibille *et al.* 2018, 4). In their discussion of the reused glass tesserae, the team noted an anecdote from al-Tabari that Umayyad caliphs had requested glass as well as craftsmen from Constantinople and suggest that this is indicative of practices that existed at least until imperial Samarra. However, the use of newly commissioned tesserae makes clear that new material was available. Again, we can see that at the most elite level in Samarra, circularity was practiced.

Other decorative items

An extract in *Kitāb al-Hadāyā wa al-Tuḥaf* describes an ambassador's visit to Baghdad in AD 917, which likely also reflects court practice in Samarra (this text is also known as *Book of Gifts and Rarities* by Ibn al-Zubayr, a collection of anecdotes considered to have been finally put together in the 11th century, and whose earliest manuscripts date to the 15th century; see discussion in Qaddumi 1996). The anecdote details priceless quantities of equipment: 'Five thousand gilded coats of mail, ... about ten thousand leather shields, helmets, [egg-shaped] helmets, plate armour, mail armour, ornate quivers, and bows' decorating the halls. These items may have been taken as booty and displayed out of war time or received as part of a gift-exchange with Constantinople; gift exchange 'was a significant aspect of medieval Islamic court society' (Anderson 2009, 110). Whatever their origin, decoration was not their primary purpose, so this group of materials could be classed as secondary use. They were also relocated from palace to palace, city to city, like the textiles and other decorative elements. The coats of mail were gilded, or modified, to allow for a permanent re-employment as decoration.

The same text informs us of a number of other examples of precious, inherited items that were used to display Abbasid power. A fragment of a mirror understood to have been God's gift to Adam, gifted to the Umayyad caliph al-Muawiyah, had been part of the Umayyad caliphal treasury until it came to the Abbasids, for instance, and was on display at Samarra (Qaddumi 1996, 174–176 para. 202). It is possible that we should understand these items as something slightly closer to relics than heirlooms. Other examples of precious items that were passed on to new users within the caliphal line include the Great Pearl of the Umayyads, a Sasanian ruby ring stone and a Sasanian bejewelled gold brazier (see Hoffman 2008 for more discussion on these artefacts, their lineage and likely new meanings).

Conclusion

This is only a brief overview of examples of reuse in imperial Samarra under the Abbasid caliphs, and future work could explore many of the groups of materials surveyed here in more detail. Research on recycling in the markets of the jewellers, swordsmiths and moneychangers would also be fruitful, as well as, for example, the role of booty in the economy from the perspective of the recipients' usages. Samarra's military cantonments and imperial mint would make it a fascinating case study for more circular economic research, particularly with regards to metals from the perspective of multiple different agents.

Even this introductory survey shows, however, that there were elements of economic circularity being practiced by the Abbasid caliphs in their palaces at imperial Samarra: new resources were frequently not consumed where there was an alternative, and so waste was avoided, whether in building or decorative materials. It is an introductory step showing the possibilities of reuse research at Samarra, demonstrating that considering economic circularity in such a prestigious, wealthy context allows us to think differently about economic systems and adds complexity to our understanding of ancient circularity in urban environments. When a circular perspective is applied to the existing literary and material evidence, it is clear that the caliphs practiced reuse (including relocation, later cycling and secondary use), recycling and conservatory processes at their imperial capital. There is no evidence that the caliphs reduced waste and avoided using new resources when existing resources were available because of a sustainable intent, as we might understand it in the present day. Rather, circular practices were one default aspect of elite economic behaviour at imperial Samarra. This further demonstrates the need to think about pre-modern economies with more nuance, including those where there were no financial constraints on resources.

Bibliography

Websites

The Circular Economy and Urban Sustainability in Antiquity project: https://projects.au.dk/circulareconomy [accessed 9 February 2023]

Secondary sources

Al-Jannabi, T. (1982) Islamic archaeology in Iraq: recent excavations. *World Archaeology* 14, 305–327.

Anderson, G. (2009) Islamic spaces and diplomacy in Constantinople (tenth to thirteenth centuries CE). *Medieval Encounters* 15.1, 86–113.

Andrade, N. and Raja, R. (2022) Alternative urban economies: the case of Roman Palmyra. *Journal of Urban Archaeology* 6, 31–47.

Awwad, G. (2008) *The Shabushti's Book of Monasteries: Al-Diyarat*. Piscataway, Gorgias Press.

Barfod, G.H., Freestone, I.C., Lichtenberger, A., Raja, R. and Schwarzer, H. (2018) Geochemistry of Byzantine and early Islamic glass from Jerash, Jordan: typology, recycling and provenance. *Geoarchaeology* 33, 623–640.

Barfod, G.H., Freestone, I.C., Jackson-Tal, R.E., Lichtenberger, A. and Raja, R. (2022) Exotic glass types and the intensity of recycling in the northwest Quarter of Gerasa (Jerash, Jordan). *Journal of Archaeological Science* 140, 1–18.

Bavuso, I., Furlan, G., Intagliata, E.E. and Steding, J. (2023) Circular economy in Roman and early medieval Europe – methods of analysis for a future agenda. *Open Archaeology* 9(1), 20220301. https://doi.org/10.1515/opar-2022-0301

Birch, T., Orfanou, V., Lichtenberger, A., Raja, R., Barfod, G., Lesher, C.E., Schulze, I. and Schulze, W. (2019) From nummi minimi to fulūs—small change and wider issues: characterising coinage from Gerasa/Jerash (Late Roman to Umayyad periods). *Archaeological and Anthropological Sciences* 11, 5359–5376.

Boschetti, C., Lichtenberger, A., Raja, R., Wootton, W., and Schibille, N. (2021) Loose glass tesserae and lost decorations: chronology and production of mosaics from Gerasa's Northwest Quarter. *Archaeometry* 63.5, 960–974.

Brown, P.J. (2022) The 'Cup of Pharaoh' from Samarra and the reuse of ancient spolia as water features in the medieval Islamic World. *Journal of Islamic Archaeology* 9.1, 59–81.

Carvajal López, J.C. (2022) Something ends, something begins: refuse and reuse in Islamic archaeology. *Journal of Islamic Archaeology* 9.1, 1–5.

Collins, B.A. (1994) *Al-Muqaddasī: The Best Divisions for Knowledge of the Regions: Aḥsan al-Taqāsīm fī Ma'rifat al-Aqālīm*. Reading, Garnet Publishing.

Dahmani, F. (2021) ʿAn al-ḥawḍ al-ḥajarī bi-l-madrasa al-sharābiyya bi-Baghdād: iʿādat naẓar fī tārīkhihi wa-fī taʿūfihi bi-qaṣʿat firʿawn. *Al-Abhath* 69, 211–243.

Dospěl Williams, E. (2019) A taste for textiles. *Dumbarton Oaks Papers* 73, 409–432.

Duckworth, C. N. and Wilson A. (2020a) Introduction. Recycling and reuse in the Roman economy. In C.N. Duckworth and A. Wilson (eds) *Recycling and Reuse in the Roman Economy*, 1–8. Oxford, Oxford University Press.

Duckworth, C.N. and Wilson, A. (eds) (2020b) *Recycling and Reuse in the Roman Economy*. Oxford, Oxford University Press.

Dumser, E. A. (2018) Visual literacy and reuse in the architecture of late imperial Rome. In D. Ng and M. Swetnam-Burland (eds) *Reuse and Renovation in Roman Material Culture: Functions, Aesthetics, Interpretations,* 140–159. Cambridge, Cambridge University Press.

Fishbein, M., Gordon, M.S., Robinson, C.F. and Rowson, E.K. (eds) (2018) *The Works of Ibn Wāḍiḥ al-Yaʿqūbī*: an English Translation *(3 Volumes).* Leiden, Brill.

Furlan, G. (2023) Aspects of circular economy on an urban and inter-urban scale in the Roman age: towards a more holistic model. *Thiasos* 12, 341–366.

Furlan, G. and Andreatta, C. (2023) Waste Nothing: The Impact of Glass and Metal Recycling in Imperial Roman Towns. *European Journal of Archaeology*, 1–19. doi:10.1017/eaa.2023.25

Goitein, S.D. (2020) *A Mediterranean Society: The Jewish Communities of the Arab World as Portrayed in the Documents of the Cairo Geniza*. Berkeley, University of California Press.

Gonnella, J. (2010) Columns and hieroglyphs: magic 'spolia' in medieval Islamic architecture of northern Syria. *Muqarnas* 27, 103–120.

Heidemann, S. (2006) The Citadel of al-Raqqa and fortifications in the Middle Euphrates area. In H. Kennedy (ed.) *Muslim Military Architecture in Greater Syria*, 122–150. Leiden, Brill.

Hoffman, E.R. (2008) Between East and West: The wall paintings of Samarra and the Construction of Abbasid princely culture. *Muqarnas Online* 25.1, 107–132.

Intagliata, E.E. (2021) Reuse of building material and sculpture in late antique and early Islamic Palmyra (273–750 CE). An overview of the practice and several remarks on the evidence from the Sanctuary of Baalshamin. *Syria* 98, 419–438.

Johnson, S.C. (2017) 'Return to origin is non-existence': Al-Mada'in and perceptions of ruins in Abbasid Iraq. *International Journal of Islamic Architecture* 6.2, 257–283.

Kadoi, Y. (2014) The Samarra Finds in the New World: Fragments of 'Abbasid Artistic Legacy in North American Museums. In J. Gonnella, S. Struth and R. Abdellatif (eds), *Beiträge zur Islamische Kunst und Archäologie*, iv, 28–37. Wiesbaden, Reichert.

Kennedy, H. (1990) *The History of al-Ṭabarī, vol. XXIX: Al-Manṣūr and al-Mahdi.* Albany, State University of New York Press.

Kennedy, H. (2005) *When Baghdad Ruled the Muslim World: The Rise and Fall of Islam's Greatest Dynasty.* Cambridge, De Capo Press.

Lassner, J. (1970) *The Topography of Baghdad in the Early Middle Ages: Text and Studies.* Detroit, Wayne State University Press.

Lassner, J. (1980) *The Shaping of Abbasid Rule.* Princeton, Princeton University Press.

Mackie, L. (2015) *Symbols of Power: Luxury Textiles from Islamic Lands, 7th–21st Century*. New Haven, Yale University Press.

Meinecke, M. (1995) al-Raḳḳa. In C.E. Bosworth, E. van Donzel, W.P. Heinrichs and G. Lecomte (eds) *The Encyclopedia of Islam, New Edition, Volume VIII*, 410–414. Leiden, Brill.

Meisami, J.S. (2001) The palace-complex as emblem: zome Samarran *Qaṣīdas*. In C. Robinson (ed.) *A Medieval Islamic City Reconsidered. An Interdisciplinary Approach to Samarra*, 69–79. Oxford, Oxford University Press.

Milwright, M. (2001) Fixtures and fittings: the role of decoration in Abbasid palace design. In C. Robinson (ed.) *A Medieval Islamic City Reconsidered. An Interdisciplinary Approach to Samarra*, 79–109. Oxford, Oxford University Press.

Milwright, M. (2017) Samarra and Abbasid ornament. In G. Neçipoğlu and F.B. Flood (eds) *The Companion to Islamic Art and Architecture*, 177–196. Hoboken, Wiley-Blackwell.

Mulder, S. (2022) Introduction. In S. Mulder (ed.) *Imagining Antiquity in Islamic Societies*, 3–32. Bristol, Intellect Books.

Needham, S.P., Leese, M.N., Hook, D.R. and Hughes, M.J. (1989) Developments in the early Bronze Age metallurgy of southern Britain. *World Archaeology* 20.3, 383–402.

Northedge, A. (2001) The palaces of the Abbasids at Samarra. In C. Robinson (ed.), *A Medieval Islamic City Reconsidered. An Interdisciplinary Approach to Samarra*, 29–67. Oxford, Oxford University Press.

Northedge, A. (2005) *The Historical Topography of Samarra*. London, British School of Archaeology in Iraq.

Orfanou, V., Birch, T., Lichtenberger, A., Raja, R., Barfod, G., Lesher, C.E. and Eger, C. (2020) Copper-based metalwork in Roman to early Islamic Jerash (Jordan): Insights into production and recycling through alloy compositions and lead isotopes. *Journal of Archaeological Science: Reports* 33, 1–15.

Peña, J.T. (2007) *Roman Pottery in the Archaeological Record*. Cambridge, Cambridge University Press.

Qaddumi, G.H. (1996) *Book of Gifts and Rarities (Kitāb al-Hadāyā wa al-Tuḥaf)*. Cambridge, Harvard University Press.

Saba, M.D. (2015) A Restricted Gaze: The Ornament of the Main Caliphal Palace of Samarra. *Muqarnas* 32, 155–195.

Saba, M.D. (2022) *Impermanent Monuments, Lasting Legacies: The Dār al-Khilāfa of Samarra and Palace Building in Early Abbasid Iraq*. Wiesbaden, Reichert Verlag.

Schibille, N., Meek, A., Wypyski, M.T., Kröger, J., Rosser-Owen, M. and Wade Haddon, R. (2018) The glass walls of Samarra (Iraq): ninth-century Abbasid glass production and imports. *PLoS ONE* 13.8: e0201749. https://doi.org/10.1371/journal.pone.0201749

Wendell, C. (1971) Baghdad: Imago Mundi, and Other Foundation-Lore. *International Journal of Middle East Studies* 2(2), 99–128.

Wild, J.P. (2020) The reuse of textiles in the Roman world. In C.N. Duckworth and A. Wilson (eds) *Recycling and Reuse in the Roman Economy*, 61–88. Oxford, Oxford University Press.

Winans, K., Kendall, A. and Deng, H. (2017) The history and current applications of the circular economy concept. *Renewable and Sustainable Energy Reviews* 86, 825–833.

Chapter 4

Through glass: recycling and reuse practices brought out by archaeometry and history

Line Van Wersch and Alexis Wilkin

Introduction

The study of glass and its uses offers many interesting insights into recycling practices during antiquity and the Middle Ages (Wilkin 2019). It also raises some difficulties and forces us to remain cautious in our interpretations. This chapter will begin with a brief review of the techniques involved in glass making. Then, we aim to review the study of glass for understanding 'invisible circularities.' This field has developed impressively over the last 20 years. Archaeology and archaeometry have made it possible to identify innovative avenues for researching practices that are poorly documented in the texts. The second section of the article shows this potential by presenting case studies collected at the site of Germigny-des-Prés, France. This aristocratic site of great interest offers an abundant corpus of reused and recycled glass tesserae. We will also briefly discuss examples of beads, vessels and stained glass within the context of reuse. We then will conclude by raising some (unanswered) questions about the circuits of raw materials.

Production, recycling and reuse of glass

First, before developing our view on recycling and its place in glass production, it is necessary to make an intellectual distinction between the three terms used: (1) *glass production* refers to both the manufacture of the raw material and the making of the objects. The process of late antique and early medieval glass production was divided into two main stages: first, in the primary workshops, raw materials were transformed into glass. This 'raw' or primary glass was then sent to secondary workshops where it was melted, possibly mixed and shaped (Foy *et al.* 2003). Both of these steps could include the recycling of previous materials. (2) *Glass recycling* involves a radical transformation of the objects or waste materials that were then used as raw material. This includes the remelting of glass to create new objects. (3) *Reuse of glass* involves an old object being put to a new use in a new context, without significant transformation. Examples of this include the transformation of an old intaglio into a medallion or the insertion of an old piece of flat glass, slightly altered, into a new stained-glass window.

In the first stage of production, the manufacture of glass requires silica, which is the vitrifying element in most archaeological glass dating from Late Antiquity to the early Middle Ages. During this period, silica came from sand that also provided alumina and calcium for stabilisers. To lower the melting point of silica (*c.* 1800°C), glassmakers added a flux. Three types of fluxes are attested for ancient glass: soda ash, a mineral that comes from natron, sodium plant ash that can be supplied by halophytic plants, and potash that comes from the ashes of continental plant (Moretti and Hreglich 2013, 24, 29–30). In different periods and geographical areas, these materials and recipes changed and, therefore, can be very useful in studying the mechanisms of glass production and diffusion. In this sense, the early Middle Ages are a crucial period.

Before the end of the 8th century, glass found in northwestern Europe was made from a mixture of sand and natron. From the 1st century AD, natron, a material not available in northwestern Europe, was extracted in the eastern Mediterranean. Most of the natron came from northern Egypt (Wadi-Natrun lakes, 100 km northeast of Cairo, and the Al-Barnuj region, west of the Nile Delta; Moretti and Hreglich 2013, 29). The natron was fused with sand in two main areas: Egypt and the Syrian-Palestinian coast. The primary glass from these two production areas can be distinguished by their chemical composition and the elements that are introduced from the sand. These two 'glass families' evolved over time, and specific compositions seem to be characteristic of specific periods (Schibille 2022). It is assumed that for some of these groups, such as the 'Foy 2' group (Foy *et al.* 2003), recycling had already taken place in the primary workshop(s) (Schibille 2022, 41).

The large primary glass slabs produced in tank furnaces, such as those found at Bet Eli'ezer (Gorin-Rosen 1995; 2000), were divided into pieces that were exported throughout the Roman Empire, including western Europe. They were sent to secondary workshops in different locations. Despite claims that primary workshops already existed in Germany in the late Roman period, it has so far been impossible to prove that any were active before the Middle Ages (Kronz 2022, 233). For the period under consideration, all known production sites in western Europe are secondary workshops. In these locations, the raw glass was melted, sometimes coloured, and shaped (Foy *et al.* 2003). In these production sites there is also archaeological evidence that recycling took place (Peters and Fontaine-Hodiamont 2005). From chemical analyses, this practice seems to have increased at the end of the Merovingian period (Freestone *et al.* 2008; Velde and Motteau 2013; Van Wersch *et al.* 2017).

It is assumed that the practice of recycling was stimulated mainly by the shortage of raw oriental glass. The scarcity of available fresh raw materials (Shortland *et al.* 2006) certainly intensified the practice of recycling as a response to production/distribution failures. It should be noted that the idea of a disruption of trade between the eastern Mediterranean and the Merovingian world is now much debated. Some scholars believe that trade was sustained until the beginning of the 8th century, a period when Merovingian burial deposits were interrupted, thus making the quantification of trade flows much more difficult (Drauschke 2019). To suggest that recycling or reuse only takes place in a context of scarcity is also to forget that the practice of recycling has always existed and is found in almost all ancient contexts. In the Roman Empire, even when raw primary glass was still in circulation, the collection and trade of broken glass is a documented practice. Cullet was found as cargo in the holds of ships, as evidenced by wrecks such as the Felix Iulia (Silvestri 2008). The addition of broken glass to new glass blocks reduced the cost of production by lowering the melting temperature and, of course, by saving on imported raw materials.

Glass recycling has been the subject of several papers by leading researchers (Freestone 2015; Grünewald and Hartmann 2015; Paynter and Jackson 2016). As Ian Freestone wrote in 2015, the chemical markers of this practice are now quite clear. Increases in three different types of 'contaminants' can reveal recycling: (1) colouring or opacifying elements. Even though glass was sorted by colour, small coloured fragments may have been introduced into the batches, bringing with them more copper, tin, zinc, cobalt, antimony or lead. (2) Elements from the smoke and ash of the fire. The increase in potassium, magnesium and phosphorus is assumed to be due to the firing process. (3) Elements from working tools. Their contact with the glass will mainly lead to an increase in the iron content. However, it is very complicated and probably impossible to estimate the amount of fresh and recycled glass in an object or to say how many times glass has been remelted, as many variables need to be taken into account. The care taken by the craftsmen in sorting and selecting materials for recycling may have been random. Furthermore, even in a batch from the same crucible, the elements are not always homogeneous, and the 'pollution' may be localised in specific areas.

In addition to these transformations due to recycling, successive reheating would affect the loss of sodium, which is a light and volatile element (Freestone 2015, 30). Dirt could also be brought in during the process and cause unwanted inclusions. These could make the glass more difficult to work. The dirt had to be removed before shaping, and the loss of fluxes would have affected the viscosity of the glass. Thus, the material would gradually have become unusable for blowing. This could have led to craftsmen developing new recipes.

Although the transition from recycled natron glass to new compositions is not yet entirely clear, it occurred in the 8th century (Shortland *et al.* 2006). In the eastern Mediterranean, soda-lime glass was replaced by soda-ash glass, and in western Europe, wood-ash glass is first documented in Charlemagne's palace in Paderborn (North Rhine-Westphalia, Germany): the palace's sacking in AD 778 provides a clearly dated context (Wedepohl *et al.* 1997). Karl Hans Wedepohl's analyses revealed that most of the glass found in the early Middle Ages was soda-lime glass, while some pieces of window glass were made from wood ash. More recently, excavations at the monastery of Baume-les-Messieurs in the Jura revealed 1,574 pieces of window glass in a layer dating from the late 8th century. All of these pieces were made of wood-ash glass (Van Wersch *et al.* 2016).

The change in flux meant more than just a change in style: it involved real upheaval in glass production. Firstly, it ended the dependence on raw materials from the Middle East and the need to recycle old materials. For glassmakers, it ended any obligation to secure access to distant materials or glass waste. Secondly, instead of imported or recycled glass, large quantities of wood and suitable sand were needed to make the raw material. The study of the sand used for this local production, which seems to have been tertiary sand (Wedepohl *et al.* 2011), is currently in full swing. This raises the question of the location of workshops in relation to raw material resources and their transport. Thirdly, this change could have completely reorganised workshops since primary and secondary production could have been carried out in the same place. Thus, in addition to the change in consumption habits, the switch from soda- to wood-ash glass had a real impact on the organisation of the craft.

While the decrease in imports explains this change, the increasing need for glass is also a factor. For the early Middle Ages in northwestern continental Europe, glass was used mainly in three types of objects: beads, vessels and architectural glass. The former and the latter are known mainly from Merovingian tombs and seem to be less numerous in the late Merovingian period. At least, we have less evidence for them. On the other hand, with the revival of monumental architecture, especially in Carolingian territories, the demand for architectural glass was likely to have increased, and while the quantities needed for small objects may have been quite reasonable, those required for window panels and mosaics may have been much larger. In the following sections, we will give concrete examples of reuse and recycling for beads, vessels and architectural glass, with particular focus on archaeology and archaeometry.

Beads

During the Merovingian period, beads, especially glass beads, became very popular (Pion 2021). These small coloured objects are found in large quantities in most female graves (Pl. 2). They seem common and accessible to the general population. The small objects were both imported from far away regions and produced locally; however, Indian beads were not found after the first quarter of the 6th century. And, from the second half of the 7th century onwards, the supply of beads from the Near East seems to have stopped. Yet, production in northwestern Europe reached its peak at this time (Pion and Gratuze 2016).

As demonstrated by Boschetti (this volume, Ch. 5), the reuse of ancient beads and pendants is attested by several necklaces and bracelets found in graves (Pion 2012). It is not uncommon to find melon beads from the Roman period or pieces of Celtic bracelets isolated or strung on a wire as part of a necklace or bracelet.

On the other hand, while recycling of beads is rather complicated to prove, recycling for the production of beads is attested by several discoveries. In Scandinavia and the present-day Netherlands, tesserae have been found in bead workshops with some examples inlaid in crucibles, partially melted (Sablerolles *et al.* 1997; Sablerolles and Henderson 2012). These tesserae were apparently used for their material, glass, and not to add colour. Indeed, melting red, yellow and white glass would result in a loss of colour (Crocco *et al.* 2021). At the same time, as evidenced by studies of beads from the Netherlands and Belgium, new recipes appeared in this period, probably due to the lack of material, especially for the popular black, red and yellow beads (Mathis *et al.* 2022; Langbroek *et al.* forthcoming). Including large amounts of metal oxides, they place bead production on the border between glass and metal crafts. In contrast, the production of dark blue beads seems to have relied mainly on recycling, as is the case for other glass objects of this colour (Gratuze *at al.* 2018).

Glass vessels

As glass vessels, common in Merovingian burials, disappeared from cemeteries, their evolution becomes more difficult to follow. However, this change was not sudden; a decrease in the number of (glass) objects in tombs can be observed from the 6th century onwards. At the same time, the shapes of these vessels became simpler (Van Wersch 2011). The most common Carolingian vessels are funnel beakers, cylindrical or globular beakers made of pale greenish or blue-green glass. But there are also, in smaller quantities, globular bowls and beakers made of highly coloured glass, decorated with *reticella* or gold rods (Rodrigues and Brut 2017, 39–40).

Bowls and blue-green glass (re)appeared in the middle of the 7th century, such as those found in the Vrijthof cemetery in Maastricht (Pl. 3). The walls of the vessels are thicker, as if the glass were less fusible. The material also appears to be full of bubbles, lines and impurities. Its composition shows clear traces of recycling, as well as colouring metal oxides elements from the ashes. It is also clear that this material was no longer discoloured by the addition of manganese (Van Wersch *et al.* 2017).

Architectural glass: window glass and tesserae

Architectural glass seems to have become more important from the middle of the 7th century onwards, due to the revival of monumental architecture. Unfortunately, no window glass from this period is still *in situ* today. As far as we know, there is no evidence for the reuse of glass pieces, as was the case in the following periods. However, some of the collections analysed have proven the recycling of glass for the production of windows or other uses. Indeed, an exceptional testimony by Gregory of Tours tells us of the theft of window glass from a church to feed the clandestine resale circuit (see below).

In most naturally coloured soda glasses, the chemical elements characteristic of recycling (colouring elements: lead, copper, zinc, tin, antimony) are present in large quantities (Van Wersch *et al.* 2014, 229; Van Wersch *et al.* 2016, 6). Similarly, deep blue glass, so-called 'cobalt blue' glass, has a composition characteristic of recycling. For example, at Stavelot (Pl. 4), as at many other sites, these fragments still have a typical natron glass composition, whereas other coloured glass is wood-ash glass. Furthermore, the cobalt signature indicates the recycling of Roman glass, and the presence of antimony marks the melting of cobalt blue Roman tesserae for this purpose (Van Wersch *et al.* 2014, 230–231). These characteristics continued for a long time, as the production of blue windows relied on the recycling of Roman tesserae until the 13th century (Gratuze *et al.* 1992, 106).

The circulation of glass tesserae is documented by the famous episode of the palace of Theodoric of Ravenna (see below). In Germigny-des-Prés (Pl. 5), glass tesserae testify to both reuse, as defined above, and recycling. For dark blue tesserae, the cobalt/nickel ratio is characteristic of the recycling of glass made before and after the 4th century. The presence of antimony also supports this hypothesis (Van Wersch *et al.* 2019, 19). It may be that this represents either recycled glass that was melted and shaped into tesserae or reused tesserae from the Roman period. Reuse of earlier tesserae is also suspected for other colours, as antimony was detected in some of them (Van Wersch *et al.* 2019, 20). In almost all coloured and opacified tesserae, we can observe chemical variations in examples of the same colour. We therefore assume that they may have come from different workshops. The recycling and reuse of old Roman tesserae may also explain this chemical diversity.

In contrast, gold- and silver-leafed tesserae made of blue-green glass appear to have been a homogeneous production. The glass compositions correspond to a common Egyptian glass group of this period. However, in the glass of the base of the tesserae, the manganese is lower than in the 'Foy 2' group, whereas it is higher in the cartellina of the tesserae. Still, in the glass of the cartellina, the elements from recycling, notably the metal oxides, copper, lead, cobalt, zinc, as well as potassium and phosphorus, are lower than in the glass of the base. These elements clearly distinguish the two parts of the tesserae (Van Wersch *et al.* 2019, 22–23): the base was made from recycled glass, while to cover the metal sheet but let it appear, the cartellina was made with much cleaner, uncoloured and carefully selected glass.

The presence of antimony in the cartellina was of the same level of concentration as in the base glass. Furthermore, the cartellina was composed of 1% manganese, which was clearly added voluntarily by the craftsmen who made the tesserae or by earlier glassmakers. In this case, the presence of manganese would also indicate a very careful selection of uncoloured glass. The producers of these tesserae therefore appear to have had access to different materials and/or techniques than other glass craftsmen, as manganese was very rarely used to discolour contemporary window glass and glass vessels.

Final thoughts on reuse and recycling. Circuits, benefits and issues of scale

The examples considered for beads, tableware and architectural glass, and the initial reflections on the conditions of glass production, speak for themselves. They demonstrate the intensity of reuse and recycling practices, even in contexts where raw materials were not in short supply. In Roman antiquity, it seems that recycling was a common practice, at least in urban areas, and that cullet was not of great economic value: peddlers collected broken vessels from door to door, exchanging them for sulphur matches, for example (Whitehouse 1999). But this is an urban context, in a commercial setting, where glass was present in abundance. A fortiori, recycling practices were more intense when such resources were lacking. Moreover, recycling was advantageous for technical reasons – e.g., the colouring of glass – and for economic reasons – it requires lower temperatures. In this case, recycling saved resources such as wood; this resource was fundamental, as in all other pyrotechnologies, and certainly explains the location of some workshops in wooded areas (*e.g.*, the site of Kordel in the Hochwald, near Trier: Janssen 1983, 323–325).

What were the sources of supply for reused glass (especially tesserae) or recycled glass in the early Middle Ages? It is very likely that surviving ancient monuments were an important source, as shown by the case of the villa at Aiano in Italy (Boschetti 2021). Given the weight of the glass, it can also be suggested that local sources of supply were favoured and that transport may have been by road, on the backs of animals or by cart, for small quantities or by river or sea, if the distances to be covered were long and the quantities larger (Loveluck 2013, 55).

In the study of supply chains, exceptional situations should not be taken as a model, such as the famous transport of tesserae and columns from Ravenna to Aachen. These endeavours were supported by Pope Hadrian I, who gave a new horse to Charlemagne after the first horse had died undertaking the transport (see MGH *Epistolae* 3, no. 81, 614; *c.* AD 787). The importance of the site at Ravenna may justify a form of 'symbolic translation of power' through the circulation of materials. The scale of the Aachen site could also pragmatically explain the need for raw materials; there is evidence of the on-site manufacture of gold leaf tesserae at this site (see Giertz and Ristow 2013).

However, the trafficking in crushed glass also existed on a smaller scale. There is at least one exceptional document by Gregory of Tours, who, in his *Liber in Gloria martyrum* (ch. 58, 78), informs us of the nightly theft of church windows for remelting, an activity that seems to have been lucrative. This text clearly indicates that glass was a valuable commodity (less so than lead, however, which was the thief's primary goal). This commodity may have been exported over distances, in the form of cullet.

In a period when the circulation of raw oriental glass might have been reduced, ground glass was sometimes a mobile commodity. It was also likely to have been exchanged within aristocratic networks, such as those linking the great Merovingian and Carolingian abbots to the lay elites who circulated the resources they sought by correspondence. In some cases, these letters show us that important individuals sent not only raw materials but also specialist technicians and glaziers (*cf.* for example Bonifatius (Wynfrith) and Lullus Moguntinus archiepiscopus, in MGH *Epistolae* 1, N. 116, 251, line 28). Sometimes glass plates were brought into circulation, for example an exchange of AD 1005 between Abbot Peringer and his bishop Gottschalk of Freising (MGH *Epistolae* 3, no. 80 and 81, 88). These were 200 *tabellas vitri* made by the abbot's glaziers; it is difficult to know their dimensions or even whether they were to be reworked or transformed by the bishop's craftsmen on his site. However, the abbey was unable to meet the demand, as it was oppressed by famine.

Alongside the circuits of exchange that connected elite places and monasteries or urban and rural centres, there was also a circularity that is almost invisible, because it is little documented in the texts and difficult to analyse by archaeometry. It is the circularity that implies an almost 'eternal' life for some objects, which were constantly and locally recast. We know, for example, from an extract from the inventory of the abbey of Tegernsee, at the head of the *Brevium Exempla* (MGH *Legum Sectio II*, no. 128, ch. 6, 250–256), that the abbey possessed, in addition to liturgical glassware, two barrels of indeterminate capacity filled with glass shards. These were stored together with other materials, such as lead and products used by the scriptorium. This was a modest quantity, probably enough to deal with the most pressing needs: repairing a broken stained glass window, for example. It was probably local waste, such as remelted objects which were added to the small quantities of cullet from local markets. These were processed by itinerant craftsmen or by specialists attached to the monasteries, who probably also provided their skills for the local population.

This example reminds us of another case observed in Jordan in the 8th century: in a context of decreasing trade in raw glass, glass hoarding was practised at the monastery site of Jabal Harum, 5 km from Petra. This same site also contained deposits of broken liturgical glass. As seen elsewhere, this liturgical glassware was excluded from the recycling circuit because of its sacred value (see Keller 2005, 65–78). Several texts, mainly Merovingian, describe in detail the accidental breaking of sacred ampullae – a significant event (see *e.g.*, MGH *Scriptores*, 351 or the Life of Caesarius of Arles, no. 51, 218–220). There were thus different categories of objects that were put back into circulation: objects of high symbolic value used, for example, for reuse (as in Ravenna/Aachen), those of high sacred and liturgical value that were removed from the recycling circuit, and those that circulated in an 'everyday' way.

Finally, the history of recycling practices must also take into account the location of the glassmakers' installations: the furnace was large, difficult to handle and had to be maintained regularly to remain in good condition. We do not know the density of the territory covered by glassmakers' kilns: in the Merovingian period in northwestern Europe, they were found in urban sites, for example on the Meuse, in Huy or in Maastricht (see Wilkin 2019). Later, they seem to have been located particularly near monasteries and elite aristocratic sites. Some communities had glazier operations in their rural domains, such as Saint-Amand-les-Eaux in AD 864 (Tessier 1943–54, vol. 2, 114), which had glaziers holding pieces of land (*mansi*). Their facilities must also have been accessible to peasant communities. In these cases, it was the people, the objects to be melted down and the cullet that had to move to the furnaces.

What was the intensity of recycling in the rural environment of the early Middle Ages? Our sources are unfortunately limited in this respect. This daily recycling, perhaps the most significant in terms of practices, largely escapes our documentary investigations but must have played an important role in everyday life.

Bibliography

Abbreviations

MGH = *Monumenta Germaniae Historica*

Primary sources

Capitularia regum Francorum I. Edited by Alfred Boretius, MGH, *Legum Sectio* II. Hanover, Hahn, 1883.

Codex Carolinus. Edited by Wilhelm Gundlach, MGH, *Epistolae* 3, 469–657. Berlin, Weidmann, 1892.

Die Briefe des heiligen Bonifatius und Lull. Edited by Michael Tangl, MGH, *Epistolae selectae,* 1. Berlin, Weidmann, 1916.

Die Tegernseer Briefsammlung. Edited by Karl Strecker, MGH, *Epistolae selectae,* 3. Berlin, Weidmann, 1925.

Gregory of Tours. *Liber in Gloria martyrum.* Edited by Bruno Krusch, MGH, *Scriptores rerum Merovingicarum* I, 34–111. Hanover, Hahn, 1885.

Morin, G., Delage M.-J. and Heijmans, M. (ed.) *Vie de Césaire d'Arles.* Paris, Editions du Cerf, 2010.

Tessier, G. (ed.) *Recueil des actes de Charles le Chauve,* 3 vol. Paris, Imprimerie nationale, 1943–1954.

Venantius Fortunatus. *Vita S. Martini.* Edited by Friedrich Leo, *MGH Scriptores, Auctores Antiquissimi,* 4.1, 293–370. Berlin, Weidmann, 1881.

Secondary sources

Boschetti, C., Gratuze, B., Cavalieri, M., Lenzi, S. and Schibille, N. (2021) Production or consumption? Glass beads from the Roman villa of Aiano-Torraccia di Chiusi, Tuscany. In: *European Journal of Archaeology*, http://hdl.handle.net/2078.1/237137 -- DOI: 10.1017/eaa.2021.34.

Crocco, R., Huisman, H., Sablerolles, Y., Henderson, J., Van Os, B. and Nieuwohof, A. (2021) Hunting colours: origin and reuse of glass tesserae from the Wierum terp. *Archaeolological and Anthropological Sciences* 13, 155. https://doi.org/10.1007/s12520-021-01391-4.

Drauschke, J. (2019) Archaeological perspectives on communication and exchange between the Merovingians and the eastern Mediterranean. In S. Esders, Y. Fox, Y. Hen and L. Sardi (eds) *East and West in the Early Middle Ages. The Merovingian Kingdoms in Mediterranean Perspective*, 9–31. Cambridge, Cambridge University Press.

Foy, D., Picon, M., Vichy, M. and Thirion-Merle, V. (2003) Caractérisation des verres de la fin de l'Antiquité en Méditerranée occidentale: l'émergence de nouveaux courants commerciaux. In D. Foy and M.-D. Nenna (eds) *Echanges et commerce du verre dans le monde antique (Actes du colloque de l'AFAV, Aix-en-Provence et Marseille, 7–9 juin 2001)*, 41–85. Montagnac, Éditions Mergoil.

Freestone, I.C. (2015) The recycling and reuse of Roman glass: analytical approaches. *Journal of Glass Studies* 57, 29–40.

Freestone, I.C., Hughes, M.J. and Stapleton, C.P. (2008) The composition and production of Anglo-Saxon glass. In V. Evison (ed.) *Catalogue of Anglo-Saxon Glass in the British Museum*, 29–46. London, British Museum Press.

Giertz, W. and Ristow, S. (2013) Goltesellae und Fensterglass. Neue Untersuchungen zur Herstellung und Nutzung von Glas im Bereich der karolingerzeitlichen Pfalz Aachen. *Antike Welt* 44, 59–66.

Gorin-Rosen, Y. (1995) Hadera, Bet Eli'ezer. *Excavations and Surveys in Israel* 13, 42–43.

Gorin-Rosen, Y. (2000) The ancient glass industry in Israel – summary of the finds and new discoveries. In D. Foy and M.-D. Nenna (eds) *Echanges et commerce du verre dans le monde antique (Actes du colloque de l'AFAV, Aix-en-Provence et Marseille, 7–9 juin 2001)*, 49–63. Montagnac, Éditions Mergoil.

Gratuze, B., Pactat, I. and Schibille, N. (2018) Changes in the signature of cobalt colorants in late antique and early Islamic glass production. *Minerals* 8, 225, https://doi.org/10.3390/min8060225.

Gratuze, B., Soulier, I., Barrandon, J.N. and Foy, D. (1992) De l'origine du cobalt dans les verres. *Revue d'Archéométrie* 16, 97–108.

Grünewald, M. and Hartmann, S. (2015) Überlegungen zum Glasrecycling der Antike im Bereich des heutigen Deutschlands. In P. Henrich, Ch. Miks, J. Obmann and M. Wieland (eds) *Non solum sed etiam. Festschrift für Thomas Fischer zum 65. Geburtstag*, 153–164. Rhaden/Westf., Verlag Marie Leidorf.

Janssen, W. (1983) Gewerbliche Produktion des Mittelalters als Wirtschaftsfaktor im ländlichen Raum. In H. Jankuhn, H. Tiefenbach, W. Janssen and R. Schmidt-Wiegand (eds) *Das Handwerk in vor- und frühgeschichtlicher Zeit*, vol. II, 331–347. Göttingen, Vandenhoeck und Ruprecht.

Keller, D. (2005) Social and economic aspects of glass recycling. In J. Bruhn, B. Croxford and D. Grigoropoulos (eds) *TRAC 2004, Proceedings of the Fourteenth Annual Theoretical Roman Archaeology Conference*, 65–78. Oxford, Oxbow Books.

Kronz, A. (2022) Rhenish glass – The role of late antique to Early medieval glass production in the rhineland. In M. Dodt, A. Kronz, B.S. Majchczack and S. Messal (eds) *Glas als Fernhandelsprodukt im frühen Mittelalter – Köln und der europäische Norden*, 233–261. Mainz am Rhein, Römisch-German Zentralmuseum Verlag.

Langbroek, M., Van Ham - Meert, A., Bordes, S., Hendriks, J., Gratuze, B., Strivay, D., Van Wersch, L. and Theuws, T. (forthcoming) Bead carnaval: big data analyses of Merovingian beads. Beads from the cemetery of Lent-Lentseveld as a case study. *Zeitschrift* für *Archäologie des Mittelalter.*

Loveluck, C. (2013) *Northwest Europe in the Early Middle Ages, c. AD 600–1050: A Comparative Archaeology*. Cambridge, Cambridge University Press.

Mathis, F., Gratuze B. and Vrielynck, O. (2022) Étude de composition et de provenance des perles en verre mérovingiennes du cimetière de Bossut-Gottechain (Belgique, Wallonie). In Y. Heningfeld and E. Peytremann (eds) *Un monde en mouvement: la circulation des hommes, des biens et des idées à l'époque merovingienne (Ve–VIIIe siècle), 40e journées internationales de l'Association française d'Archéologie mérovingienne (AFAM), Nantes: 3–5 octobre 2019*, 211–228. Saint-Germain-en-Laye, Association française d'Archéologie mérovingienne.

Moretti, C. and Hreglich, S. (2013) Raw materials, recipes and procedures used for glass making. In K. Janssens (ed.) *Modern Methods for Analysing Archaeological and Historical Glass*, 23–47. Chichester, Wiley.

Paynter, S. and Jackson, C.M. (2016) Re-used Roman rubbish: a thousand years of recycling glass. *Post-Classical Archaeologies* 6, 31–52.

Péters, C. and Fontaine-Hodiamont, C. (2005) Huy et le travail du verre à l'époque mérovingienne: étude préliminaire du matériel trouvé rue Sous-le-Château et place Saint-Séverin. In J. Plumier and M. Regnard (eds) *Voies d'eau, commerce et artisanat en Gaule mérovingienne*, 233–268. Namur, Ministère de la Région Wallonne.

Pion, C. (2012) La pratique du remploi dans les sépultures mérovingiennes de Belgique. Entre recyclage, esthétique et symbolique. *Cahier des thèmes transversaux ArScAn* 10, 47–55.

Pion, C. (2021) De verre et d'ambre. Les parures en perles du cimetière mérovingien de Ciply (B.). In M. Demelenne and G. Dumont (eds) *Le monde de Clovis, Itinéraires mérovingiens, Musée royal de Mariemont*, 205–213. Morlanwelz-Mariemont, Musée royal de Mariemont.

Pion, C. and Gratuze, B. (2016) Indo-Pacific glass beads from Indian subcontinent in early Merovingian graves (Vth–VIth century AD). *Archaeological Research in Asia* 6, 51–64.

Rodrigues, N. and Brut, C. (2017) Les balbutiements du verre de table. In S. Lagabrielle (ed.) *Le verre. Un Moyen Âge inventif, Catalogue de l'exposition présentée au musée de Cluny du 20 septembre 2017 au 8 janvier 2018*, 38–41. Paris, Editions de la Réunion de musée nationaux.

Sablerolles, Y. and Henderson, J. (2012) De glasvondsten. In M. Dijkstra (ed.) *Het domein van de boer en de ambachtsman Een opgraving op het terrein van de voormalige fruitveiling te Wijk bij Duurstede: een deel van Dorestad en de villa Wijk archeologisch onderzocht*, 293–350. Amersfoort, ADC ArcheoProjecten.

Sablerolles, Y., Henderson, J. and Dijkman, W. (1997) Early medieval glass bead making in Maastricht (Jodenstraat 30), The Netherlands. In U. Von Freeden and A. Wieczorek (eds) *Perlen Archäologie, Techniken, Analysen, Akten des Internationalen Perlensymposiums in Mannheim von 11 bis 14 November 1994*, 291–313. Bonn, Habelt.

Schibille, N. (2022) *Islamic Glass in the Making: Chronological and Geographical Dimensions*. Leuven, Leuven University Press.

Shortland, A., Schachner, L., Freestone, I.C. and Tite, M. (2006) Natron as a flux in the early vitreous materials industry: sources, beginnings and reasons for decline. *Journal of Archaeological Science* 33.4, 521–530.

Silvestri, A. (2008) The coloured glass of Iulia Felix. *Journal of Archaeological Science* 35.6, 1489–1501.

Van Wersch, L. (2011) Céramiques et verres mérovingiens dans la vallée mosane. Apports de l'archéologie et de l'archéométrie à l'histoire économique, sociale et culturelle. Unpublished thesis, Université de Liège.

Van Wersch, L., Biron, I., Neuray, B., Mathis, F., Chêne, G., Strivay, D. and Sapin. C. (2014) Les vitraux alto-médiévaux de Stavelot (Belgique). *ArchéoSciences* 38 doi:10.4000/archeosciences.4284

Van Wersch, L., Loisel, C., Mathis, F., Strivay, D. and Bully, S. (2016) Analyses of Early Medieval stained window glass from the monastery of Baume-Les-Messieurs (Jura, France). *Archaeometry* 58, 930–946.

Van Wersch, L., Biron, I., Mathis, F., Chêne, G. and Strivay, D. (2017) The scientific analyses of glass vessels. In F. Theuws and M. Kars (eds) *The Saint-Servatius Complex in Maastricht. The Vrijthof Excavations (1969–1970)*, 316–325. Bonn, Habelt.

Van Wersch, L., Kronz, A., Simon, K., Hocquet F.-P. and Strivay D. (2019) Les matériaux des mosaïques de Germigny-des-Prés, *Bulletin du centre d'études médiévales d'Auxerre | BUCEMA* [En ligne], Hors-série n° 11 | 2019, published online April 2019, accessed March 2023. https://journals.openedition.org/cem/16134

Velde, B. and Motteau, J. (2013) Glass compositions of the Merovingian period in Western Europe. In K. Janssens (ed.) *Modern Methods for Analysing Archaeological and Historical Glass*, 387–397. Chichester, Wiley.

Wedepohl, K.H., Winkelmann W. and Hartmann, G. (1997) Glasfunde aus der karolingischen Pfalz in Paderborn und die frühe Holzasche-Glasherstellung. *Ausgrabungen und Funde in Westfalen-Lippe* 9 A, 41–53.

Wedepohl, K.H., Simon, K. and Kronz, A. (2011) Data on 61 chemical elements for the characterization of three major glass compositions in Late Antiquity and the Middle Ages. *Archaeometry* 53, 81–102.

Whitehouse, D. (1999) Glass in the Epigrams of Martial. *Journal of Glass Studies* 41, 73–81.

Wilkin, A. (2019) Le travail et commerce du verre au haut Moyen Âge, un révélateur pour l'histoire économique? In D. Strivay, L. Van Wersch, and F. Theuws (eds) *Early Medieval Tesserae in Northwestern Europe (VIth–Xth century): Material, Techniques and Exchanges*, 14–25. Bonn, Habelt.

Chapter 5

Beauty, utility and value: examples of glass reuse from the Roman period to the early Middle Ages

Cristina Boschetti

Introduction

The fortune of glass in antiquity was surely due to a combination of functional and aesthetic qualities. The colour of glass was intentionally modified adding metal oxides and pigments, which could colour, opacify and even decolourise the glass, to obtain a perfectly clear product (Jackson 2005; Paynter *et al.* 2015). Strong colours in glassware were greatly appreciated during the Hellenistic period, but colourless glass started to become gradually predominant from the beginning of the 1st century AD, when glassblowing had become an established practice across the Mediterranean region (Stern 1999; Grose 2017; Larson 2019). This new technique, which began to be practiced in the Syro-Palestinian region around the middle of the 1st century BC (Israeli 1991), literally revolutionised the glass market, shortening significantly the length of time needed to fashion glass objects (Larson 2019). With the establishment of glassblowing, for the first time, objects for everyday use were fashioned in glass, and the volume of glass circulating on the market increased significantly. This major change was surely reflected by the presence of a larger volume of glass in urban waste and by an increasing need to manage it conveniently. Recycling was surely the most straightforward strategy to cope with this problem and to economise fresh glass, but it should be remembered that it was flanked by the reuse of glass objects and fragments.

During the last couple of decades, archaeological scholarship has demonstrated the growing interest in this subject for the Roman period and the Middle Ages. It is now well established that glass was collected systematically to be remelted, as documented mainly by the evidence of recycling workshops operating across the Mediterranean basin and in northern Europe (Schibille and Freestone 2013; Boschetti *et al.* 2016; Neri and Catacchio 2016; Boschetti *et al.* 2021). Recycled glass can be identified chemically measuring the chemical tracers that reflect the pollution from exposure to the fumes of the furnace and to the accidental introduction of colouring agents (Schibille and Freestone 2013; Barfod *et al.* 2018; Boschetti *et al.* 2022). While studies focused on identifying the evidence of remelting are several, studies on the different ways that glass objects and glass fragments were reused are rare. This paper focuses on the multiple lives of ancient glass, presenting case studies documenting the reuse of glass during the period from the 2nd century BC to the 8th century AD. The discussion is broken into three sections – beauty, function and value –with each dedicated to the different reasons behind the decision to give a second life to glass objects. The aim of this paper is to offer an overview of the complex dynamics that governed the circulation of glass products and of the impact made by the minor economic phenomena that are scarcely visible in the archaeological record and about which the literary sources tend to be silent.

Beauty

When and where the Romans started to systematically practice glass recycling is still unclear (Degryse 2020), but recycling was surely intensified after the spread of glassblowing and the consequent increase in the volume of glass circulating on the market (Larson 2019). In the Mediterranean region, the earliest evidence of Roman glass recycling can be dated to the second half of the 1st century AD, thanks to the identification of a few recycled compositions in vessels from Pompeii (Brill 2012). There is a complete lack of archaeological evidence for workshops devoted to glass recycling, but there is evidence that glass, even when broken, was still considered a valuable material to be used and displayed. The appreciation for the aesthetic qualities of fragments of glass vessels is reflected in the peculiar use of such fragments in the earliest examples of Roman wall mosaics, first documented in Latium at the end of the 2nd century BC (Cifarelli 2020; Pl. 6). During this early phase, floor and wall mosaics were clearly the result of two different technical traditions originating from distinct archetypes. Floor mosaics were essentially made laying tesserae – small elements cut in the form of parallelepipeds and usually made of stone – to create a flat and regular surface. On the contrary, contemporary wall decorations were uneven and multi-material surfaces, where fragments of tufa stone, seashells, pellets of Egyptian blue and even fragments of broken glass vessels were used to create the illusion of a natural rocky surface (Pl. 7).

The use of broken glass vessels in mosaics is surely a peculiar phenomenon, which has not received particular attention in scholarly literature. In the past, it has been interpreted as a practice adopted as a substitute for the more expensive and less readily available glass plaques (Sear 1977, 40; Lavagne 1988, 405–437). Although this interpretation would require an in-depth investigation, I suggest that fragmentary glass vessels might have been selected for their own nature and not as a surrogate for more desirable materials. The decorations that include glass fragments tend to be dominated by a taste for the non-geometric shape of the different elements that create the revetment. The fragments of glass are arranged unevenly, along with blocks of tufa, chips of stones and shells. The broken glass vessels were likely selected for their bright colour and for their irregular shape, which coordinate perfectly with the uneven shapes and sizes of the shells.

An analysis of the Latin sources offers some hints that push the interpretation of these fragmentary vessels used in wall mosaics beyond purely aesthetic explanations. In the Latin tradition, fragility is noted as a quality that increased the value of glass (Citroni Marchetti 2003). In the literature of the High Empire, the potential of glass vessels to break is a literary *topos*, where shattered glass becomes a reminder of the fragility of political power and beauty (Leone 2017). Considering that these wall mosaics were part of the decorative apparatus of display in public spaces of houses and villas, like *triclinia* and furnished gardens, it is possible that the fragmentary vessels were part of a game of allusion, connecting the decoration of the space to the atmosphere of the banquet. The use of fragments of glass vessels in wall mosaics ceased around the age of Tiberius, when wall-mosaic techniques changed radically, mainly as a consequence of the introduction and diffusion of glass tesserae (Boschetti 2011). After this period, the most prominent parts of wall decorations were dominated by figural, geometric or vegetal motifs. Broken glass vessels were still present, but they tended to be relegated to the less visible parts of the decoration. In two contexts, the nymphaeum excavated in Piazza Marconi at Cremona and in the House of the Grand Duke of Tuscany in Pompeii, few opaque sealing-wax red glass tesserae were cut from fragments of broken vessels (Boschetti 2011; Pl. 8). Interestingly, in these two mosaics the decoration combines areas with glass vessels, shells and tufa stone and areas in *tessellatum*. This combination of styles and materials means that the workshops had access to the palette traditionally used in the earliest wall mosaics, including fragments of vessels, as well as to the flat glass cakes used to cut the tesserae. Opaque red glass is difficult to produce, and the sealing-wax red is particularly challenging (Boschetti *et al.* 2017). It is clear that, in this case, the use of vessel fragments was clearly utilitarian and functional, in order to supply to the lack of a colour that was not easily available.

The relatively limited circulation of glass vessels in Italy during the 2nd and the 1st centuries BC suggests that the fragments used by mosaicist workshops were not casually collected on site but were supplied by a network of collection and distribution of broken glass. Interestingly, the recent investigations conducted on Iron Age glass jewellery from Britain demonstrate that, as early as the 2nd century BC, coloured glass produced in the Mediterranean region was reworked and even re-coloured locally to fashion bangles and beads (Jackson and Paynter 2022; Paynter *et al.* 2022). Considering that some of the finds analysed by Sarah Paynter and Caroline Jackson significantly predate the Roman conquest of Britain, it is not likely that this glass was supplied locally from the sorting and collection of broken vessels from general waste. In light of this recent discovery, it becomes clear that the fragments used for mosaic decorations would have been sourced from an organised system for the collection and distribution of broken glass that operated on international scale.

Utility

Cutting tesserae from fragments of vessels is not the only case in the Roman world where glass shards were used for practical purposes or adapted to create objects with a new function. Latin treatises on agricultural surveying mention the use of glass and ceramic shards as *signa* to mark the limits of properties (Leone 2017, 13). The adaptation of shards is rarely studied, but the survey conducted by Sylvia Fünfschilling (2015) on the glass finds excavated in Augusta Raurica produced a good classification of the different categories of these artefacts. Bases of cups and dishes with knapped walls and quadrangular or triangular fragments with knapped edges recur regularly, together with blades worked like flint and fragments of flat glass of regular shape. The function of these artefacts is not completely clear, but they are often identified as tools, like smoothers of knives, or decorative plaques, such as a fragment of a red glass plaque excavated in the inlay workshop operating in Denderah, Egypt, during the 1st century BC (Boschetti 2018; Pl. 9). The findspots of reworked glass fragments are usually dwellings in urban contexts, suggesting a possible connection with minor crafts practiced inside the domestic sphere (Fünfschilling 2015).

A similar reuse of glass fragments is visible in the archaeological record of Late Antiquity and the early Middle Ages when Roman glass fragments are occasionally documented in burials (Dévai 2019). These glass fragments tend to be classified under the category of 'odd objects' – a miscellanea of small finds of unclear function – and identified as objects charged with supernatural and magical power (Page and Rider 2019, 384). However, recent studies are offering a more nuanced interpretation of these objects as indicators of social status (Horvath and Bendo 2011; Boschetti *et al.* 2023). It is worth mentioning that in male burials of the Migration period, such as the tomb of the smith in the Cemetery of Campo Marchione, Italy, the glass fragments tend to be associated with sets of firesteel and flint (La Salvia 2017). This association suggests a possible connection between the glass fragments and these tools. It is possible that the broken vessels were used either to hold the starter for lighting a fire or as flints.

Interestingly, the adaptation of glass shards is not a phenomenon documented exclusively in antiquity. The modern anthropological record offers similar examples of the reuse of glass fragments for new purposes. This phenomenon has touched nearly all pre-industrial societies of the globe. The most recent studies on this subject demonstrate that previous scholarly literature has overestimated the diffusion and frequency of this practice (Becker 2022). For instance, it is now established that entire productions of glass tools, like the colourful bottle-glass arrowheads manufactured by the Aboriginals of the Kimberley district, in Northwestern Australia, were actually produced for the collectors trade and originated after the first contacts with the Europeans (Becker, 2022, 34–36; Pl. 10). Nevertheless, despite rare, knapped-glass tools or tools made with adapted glass shards are documented from the 16th century onwards in North America, South America and Africa (Becker 2022). Like in the Roman world, the adaptation of broken glass was clearly a marginal phenomenon, but it existed alongside in parallel to the main dynamics shaping the glass economy.

Value

Glass objects used for an extended period or traded and exchanged over long distances can be filtered by a process of semiotic translation. Unlike the broken vessels, these objects remain intact and are not reworked, but their final function and meaning can either be modified or become ambiguous. In other words, the value of these object can change. Glass beads, in particular, represent an excellent case study to investigate the shifting meanings of glass objects. During the Migration period and the early Middle Ages, colourful glass beads were popular body ornaments and are common finds in Europe (Koch, 1977; Guido, 1999; Burgmann, 2004), where they were frequently deposited in furnished female and, occasionally, also in male burials (O'Sullivan, 2015; Boschetti *et al.* 2020). These artifacts, scarcely considered in archaeological literature over the past few decades, are now receiving growing attention in archaeological studies. This stems from the identification of beads as important indicators of social status and markers of long-distance trade networks (Pion and Gratuze 2016; Neri *et al.* 2019; Boschetti *et al.* 2020).

The beadmaking workshops located in the northern sector of Europe have always been important actors in the circular economy of glass. As mentioned previously, the workshops active in the British islands in the production of beads and bangles were consumers of Roman glass cullet as early as the 2nd century BC (Jackson and Paynter 2022; Paynter *et al.* 2022). Later, during the early Middle Ages, glass recycling was a regular practice amongst the bead-making workshops identified in Scandinavia and northern Europe (Sablerolles *et al.* 1997; Skre 2011, 172–173, 232–238; Crocco *et al.* 2021; Maixner 2021; Barfod *et al.* 2022). The combination of typological, technological and chemical studies of glass beads retrieved from Mediterranean and Continental sites is gradually revealing the picture of a broad bead trade, where the beads manufactured in northern Europe circulated alongside beads produced in Egypt, Mesopotamia and India (Pion and Gratuze 2016; Neri *et al.* 2019; Boschetti *et al.* 2020).

During the 5th and the 6th centuries, a particular class of beads fashioned with the mosaic technique – winding and fusing together sections of polychrome glass canes – were popular items in Europe. Cylindrical or subspherical mosaic beads, pairing sections of square-mosaic bars with a white flower motif in a dark blue matrix or bars with a radiated pattern with a multilayered eye at the centre, were particularly popular. In most cases, these beads are capped at the extremities with a thread of opaque red or yellow glass (Guido 1999, 70–73; Spaer 2001, 126; Burgmann 2004, 28, 38). The connection of these capped beads to Egypt was recently ascertained thanks to their chemical composition and by the identification of identical examples in furnished late antique Egyptian burials (Brunton 1930, 27; Boschetti *et al.* 2020). This identification has confirmed the pioneering hypothesis advanced by Margaret Guido (1990, 70–73) who, in her seminal study on glass beads from Anglo-Saxon England, proposed that the bead trade had a more international dimension than was generally thought at the time.

The analysis of the Egyptian archaeological record reveals that the same combination of mosaic canes was also used to produce mosaic inlays and large – approximately 4.5 cm in diameter – circular spindle whorls, with flat bottoms and domed bodies (Silvano 2017; Pl. 11). In Egypt, mosaic glass spindle whorls were popular during Late Antiquity, alongside wound examples – manufactured by winding a small gather of hot glass or the softened tip of a glass rod around a metal wire (the mandrel) – which could be monochrome or decorated with applied threads (Spaer 2001). The relatively light weight and the small perforation of these glass spindle whorls suggests that they were used to spin light fibres, like cotton and flax (Spaer 2001). In continental Europe, the British Isles and Scandinavia, mosaic or wound glass spindle whorls tend to be interpreted as body ornaments. A recently published mosaic example from Sandy Borg, Sweden (Alfsdotter *et al.* 2018), identical to the Egyptian spindle whorls, is described as a 'sword-bead'. This definition follows the identification of these objects with rings inserted in the handle of swords, but it is now clear that this identification is not correct and these beads/whorls were attached with a string to the scabbard or to the hilt (Vallet 1988, 50). The so-called sword-beads are frequently found in furnished male graves, placed at the side of the sword, as well as in female graves, attached to the belt. Their identification in both female and male graves makes their interpretation ambiguous and does not exclude the possibility that the sword-beads were also actually used also as proper spindle whorls. What is certain, is that the meaning and function of these artefacts was fluid, and their practical function was interlaced with their symbolic value.

Whether the value of beads during the Middle Ages was monetary as well as symbolic is open to debate, but the idea that glass beads might have served as currency is becoming a mainstay in scholarship. It is difficult to find confirmation of this hypothesis in the archaeological record, but, as recently discussed by Birgit Maixner (2021, 183–185), some recurring characteristics of the archaeological contexts in which these glass beads are found support this hypothesis. For instance, loose beads are occasionally associated with silver coins in hoards of the early or the very late Viking age, when they can also be found inside purses containing coins, silver wire and precious stones. In this period, silver was scarce in the region, and the lack of metal could easily justify the greater circulation of non-monetary commodities as means of exchange. Maixner (2021, 184) also suggests that not all glass beads might have served as currency, with only specific types used by merchants for international transactions, such as paying guides, port taxes and customs duties. This possible function of beads in international trade finds a fitting parallel in the system of payments documented along the East African caravan road during the 19th century (Pallaver 2009). Here, all the non-monetary commodities used in international commerce (*e.g.*, cloth, beads and metal wire) had a strict classification and even a value that was subject to fluctuation (Pallaver 2009, 23–25).

I would also suggest that between the 4th and the 6th centuries, when the first communities of migrants from northeastern Europe settled in the Mediterranean region, glass beads could have been used for minor transactions. This possibility might seem unlikely given the area's strong Roman cultural background, but it should not be excluded outright. A similar phenomenon can be seen in Italy during the late 20th century. In 1970, the European economy was impacted by the petrol crisis. In the space of five years, the crisis produced a shortage of metals, which created a sudden lack of coins in Italy (Aresti 2015). In order to cope with this lack of currency, shopkeepers started to use different goods, including candies, biscuits and telephone tokens, as substitutes for small change. While the balance between the supply and demand for coins was finally re-established in 1978, candies continued to be used as small change for almost a decade (Aresti 2015). The reasons behind the extension of this practice beyond the period of strict necessity might seem hard to explain; however, the use of candies as a form of change had clearly become a habit, and it is likely that this habit was agreeable to customers and had a positive influence on commercial dynamics. While this might sound strange to us, I do not believe that even today anyone is really happy to leave a shop with pockets full of copper cents.

Conclusions

These reflections on the multiple lives of glass objects offer a nuanced view of the circular dimension of the glass economy during the Roman period and the Middle Ages. Glass was indeed considered beautiful, useful and valuable, and for these reasons it was not to be wasted. The collection of glass for recycling, which until a few years ago was believed to be a practice developed mainly during the Middle Ages, is now recognised as a systematic practice, at least from the early Roman Empire. The evidence for the earliest wall mosaics documents that broken glass vessels were separated from urban waste and reused to create decorations. In the same period, glass fragments circulated along international trade routes and were recycled in workshops making beads and bangles that were active in the British Isles. Their durability, hardness and transparency also made broken vessels desirable for reworking and adaptation for new and practical functions. This phenomenon is documented during the Roman period and the Middle Ages, when some broken Roman glass vessels were reworked or reused without any changes rather than entering the cycle of remelting to make new glass objects. Finally, it should be stressed that some glass objects, particularly body ornaments, were charged with a strong cultural meaning. The enormous distances covered by the ancient bead trade transformed glass beads into objects that were continuously changing in meaning, passing from hand to hand and crossing space and time.

Bibliography

Alfsdotter, C., Papmehl-Dufay, L. and Victor, H. (2018) A moment frozen in time: evidence of a late fifth-century massacre at Sandby borg, *Antiquity* 92.362, 421–436. doi: 10.15184/aqy.2018.21.

Aresti, A. (2015) 1975-Miniassegno. In *90 parole per 90 anni: 90 parole per novant'anni di invenzioni, scoperte, piccole e grandi rivoluzioni quotidiane*. Roma, Istituto della Enciclopedia Italiana fondata da Giovanni Treccani. Available at: https://www.treccani.it/catalogo/ebook/alfabeto_treccani/scienze_sociali/90_parole_90_anni.html.

Barfod, G.H., Freestone, I.C., Lichtenberger, A., Raja, R. and Schwarzer, H. (2018) Geochemistry of Byzantine and early Islamic glass from Jerash, Jordan: typology, recycling, and provenance. *Geoarchaeology* 33.6, 623–640. doi: 10.1002/gea.21684.

Barfod, G.H., Feveile, C. and Sindbæk, S.M. (2022) Splinters to splendours: from upcycled glass to Viking beads at Ribe, Denmark. *Archaeological and Anthropological Sciences* 14.9, 1-25s. doi: 10.1007/s12520-022-01646-8.

Becker, M.J. (2022) Chipped glass artifacts made digital commons by native people: cultural adaptations and innovations. *Bulletin of the Archaeological Society of New Jersey* 77, 13–58. Available at: https://digitalcommons.wcupa.edu/anthrosoc_facpub/160.

Boschetti, C. (2011) Vitreous materials in early mosaics in Italy: faience, Egyptian blue, and glass. *Journal of Glass Studies* 53, 59–91.

Boschetti, C. (2018) Working glass in Ptolemaic Egypt, a new evidence from Denderah. *Journal of Archaeological Science: Reports* 22, 550–558. doi: 10.1016/j.jasrep.2018.04.029.

Boschetti, C., Mantovani, V. and Leonelli, C. (2016) Glass coloring and recycling in Late Antiquity: a new case study from Aquileia (Italy). *Journal of Glass Studies* 58, 69–86.

Boschetti, C., Henderson, J. and Evans, J. (2017) Mosaic tesserae from Italy and the production of Mediterranean coloured glass (4th century BCE–4th century CE). Part II: Isotopic provenance. *Journal of Archaeological Science: Reports* 11, 647–657. doi: 10.1016/j.jasrep.2016.12.032.

Boschetti, C., Gratuze, B. and Schibille, N. (2020) Commercial and social significance of glass beads in migration-period Italy: the Cemetery of Campo Marchione. *Oxford Journal of Archaeology* 39.3, 319–342. doi: 10.1111/ojoa.12200.

Boschetti, C., Gratauze, B., Cavalieri, M., Lenzi, S. and Schibille, N. (2021) Production or consumption? Glass beads from the Roman Villa of Aiano, Tuscany. *European Journal of Archaeology* 25.2, 196–215. doi: 10.1017/EAA.2021.34.

Boschetti, C., Kindberg Jacobson, J., Parisi Presicce, C., Raja, R., Schibille, N. and Vitti, M. (2022) Glass in Rome during the transition from late antiquity to the early Middle Ages: materials from the Forum of Caesar. *Heritage Science* 10.1, 1–14. doi: 10.1186/s40494-022-00729-y.

Boschetti, C., Gratuze, B. and Schibille, N. (2023) Garnet trade in early medieval Europe: the Italian network. *European Journal of Archaeology* 26.1, 101–119. doi: 10.1017/eaa.2022.25.

Brill, R.H. (2012) Chemical analysis of glasses from Pompeii. In L.A. Scatozza Höricht (ed.) *L'instrumentum vitreum a Pompei*, 373–410. Roma, Aracne.

Brunton, G. (1930) *Qau and Badari III*. London, British School of Archaeology in Egypt.

Burgmann, B. (2004) *Glass Beads from Early Anglo-Saxon Graves: A Study of the Provenance and Chronology of Glass Beads from Early Anglo-Saxon Graves, Based on Visual Examination - OpenBibArt*. Oxford, Oxbow Books. Available at: http://www.openbibart.fr/item/display/10068/829434 (Accessed: 7 October 2020).

Cifarelli, F.M. (ed.) (2020) *Il ninfeo di Q. Mutius a Segni*. Roma, Quasar.

Citroni Marchetti, S. (2003) La rappresentazione del denaro in Plinio il Vecchio e nel moralismo romano. In G. Urso (ed.) *Moneta Mercanti Banchieri I precedenti greci e romani dell'Euro Atti del convegno internazionale, Cividale del Friuli, 26–28 settembre 2002*, 283–300. Pisa, Edizioni ETS.

Crocco, R., Huisman, H., Sablerolles, Y., Henderson, J., van Os, B. and Nieuwhof, A. (2021) Hunting colours: origin and reuse of glass tesserae from the Wierum terp. *Archaeological and Anthropological Sciences* 13.9, 155. doi: 10.1007/s12520-021-01391-4.

Degryse, P. (2020) Elements, isotopes, and glass recycling. In C.N. Duckworth and A. Wilson (eds) *Recycling and Reuse in the Roman Economy*, 285–300. Oxford, Oxford University Press.

Dévai, K. (2019) Re-used glass fragments from Intercisa. *Dissertationes Archaeologice* 7, 187–204.

Fünfschilling, S. (2015) The re-use of Roman glass fragments. In J. Bailey, I. Freestone and C. Jackson (ed.) *Glass of the Roman world*, 170–176. Oxford, Oxbow Books.

Grose, D.F. (2017) *The Hellenistic, Roman, and Medieval Glass from Cosa*. Ann Arbor, The University of Michigan Press.

Guido, M. (1999) *The Glass Beads of Anglo-Saxon England, c. AD 400–700*. Martlesham, Boydell Press for the Society of Antiquaries of London. doi: 10.26530/20.500.12657/50361.

Heaser, S. (2022) *Bead Furnaces in Anglo-Saxon Britain. An Investigation into the Likely Sources of Heat Used by Beadmakers in the 5th to 7th Centuries CE*. Wortham, Sue Heaser Publications.

Horvath, E. and Bendo, Z. (2011) Provenance study on a collection of loose garnets from a Gepidic period grave in northeast Hungary. *Archeometriai Műhely* 8.1, 17–32.

Israeli, Y. (1991) The invention of blowing. In M. Newby and K. Painter (eds) *Roman Glass. Two Centuries of Art and Invention* 46–55. London, The Society of Antiquaries of London.

Jackson, C.M. (2005) Making colourless glass in the Roman period. *Archaeometry* 47.4, 763–780. doi: 10.1111/j.1475-4754.2005.00231.x.

Jackson, C. and Paynter, S. (2022) Baubles, bangles and beads: Recycling coloured glasses in the British Iron Age and Roman periods. *Archaeometry* 64.S1, 150–167. doi: 10.1111/arcm.12767.

Koch, U. (1977) *Das Reihengräberfeld von Schretzheim, Bayerisch-Schwaben*. Berlin, Germanische Denkmäler der Völkerwanderungszeit.

Larson, K. A. (2019) Cheap, fast, good: the Roman glassblowing revolution reconsidered. *Journal of Roman Archaeology* 32, 7–22. doi: 10.1017/S1047759419000035.

Lavagne, H. (1988) *Operosa Antra. Recherches sur la grotte à Rome de Sylla à Hadrien*. Roma, École française de Rome. Available at: https://www.persee.fr/doc/befar_0257-4101_1988_mon_272_1.

Leone, M. (2017) On broken glass: for a semiotics of anti-materiality. *Chinese Semiotic Studies* 13.1, 73–121.

Maixner, B. (2021) Urban market production and coastal trade in the early Viking Age: a study of eighth-century AD blue period millefiori beads in central and northern Norway. *Journal of Urban Archaeology* 4, 167–191. doi: 10.1484/j.jua.5.126599.

Neri, E. and Catacchio, S. (2016) Un atelier proto-bizantino per il riciclaggio del vetro a Hierapolis di Frigia. *Orizzonti* XVII.17, 117–132. doi: 10.1400/243336.

Neri, E., Gratuze, B. and Schibille, N. (2019) The trade of glass beads in early medieval Illyricum: towards an Islamic monopoly. *Archaeological and Anthropological Sciences* 11.3, 1107–1122. doi: 10.1007/s12520-017-0583-5.

O'Sullivan, J. (2015) Strung along: re-evaluating gendered views of Viking-age beads. *Medieval Archaeology* 59.1, 73–86. doi: 10.1179/0076609715Z.00000000049.

Page, S. and Rider, C. (2019) *The Routledge History of Medieval Magic*. London, Routledge. doi: 10.4324/9781315613192.

Pallaver, K. (2009) 'A recognised currency in beads'. Glass beads as money in 19th-century East Africa: the central caravan road. In C. Eagleton, H. Fuller and J. Perkins (eds) *Money in Africa*, 1–74. London, British Museum.

Paynter, S., Kearns, T., Cool, H. and Chenery, S. (2015) Roman coloured glass in the Western provinces: the glass cakes and tesserae from West Clacton in England. *Journal of Archaeological Science* 62, 66–81. doi: 10.1016/j.jas.2015.07.006.

Paynter, S., Crew, P., Campbell, R., Hunter, F. and Jackson, C. (2022) Glass bangles in the British Isles: a study of trade, recycling and technology in the first and second centuries AD. *Antiquaries Journal* 102, 15–44. doi: 10.1017/S0003581521000378.

Pion, C. and Gratuze, B. (2016) Indo-Pacific glass beads from the Indian subcontinent in Early Merovingian graves (5th–6th century AD). *Archaeological Research in Asia* 6, 51–64. doi: 10.1016/j.ara.2016.02.005.

Sablerolles, Y., Henderson, J. and Dijkman, W. (1997) Early medieval glass bead making in Maastricht (Jodenstraat 30), the Netherlands: an archaeological and scientific investigationstraat 30), The Netherlands. An archaeological and scientific investigation. In U. Von Freeden and A. Wieczorek (eds) *Perlen: Archäologie, Techniken, Analysen: Akten des internationalen Perlensymposiums in Mannheim vom 11. bis 14. November 1994*, 293–313. Bonn, Habelt. Available at: http://www.openbibart.fr/item/display/10068/950536 (Accessed: 18 January 2021).

La Salvia, V. (2017) Il fabbro, i suoi strumenti e la sua officina. La lunga durata delle tecniche di produzione e della circolazione delle conoscenze durante l'alto medioevo. In M. Beghelli and P.M. De Marchi (eds) *I maestri del metallo, l'intelligenza nelle mani (L'Alto Medioevo. Artigiani, tecniche produttive e organizzazione manifatturiera 2. Milano, Civico Museo Archeologico, 10 Maggio 2015)*, 25–42. Roma, BraDypUS.

Schibille, N. and Freestone, I.C. (2013) Composition, production and procurement of glass at San Vincenzo al Volturno: an early medieval monastic complex in southern Italy. *PLOS ONE* 8.10: e76479. doi: 10.1371/journal.pone.0076479.

Sear, F.B. (1977) *Roman Wall and Vault Mosaics*. Heidelberg, F.H. Kerle Verlag.

Silvano, F. (2017) New finds of mosaic glass inlays from Antinoupolis, Egypt. In S. Wolf and A. De Pury-Gysel (eds) *Annales du 20e Congrès de l'Association Internationale pour l'Histoire du Verre*, 243–247. Rhaden/ Westf, Verlag Marie Leidorf.

Skre, D. (ed.) (2011) *Things from the Town. Artefacts and Inhabitants in Viking-age Kaupang (Kaupang Excavation Project Publication Series)*. Aarhus, Aarhus University Press/ the Kaupang Excavation Project, University of Oslo.

Spaer, M. (2001) *Ancient Glass in the Israel Museum. Beads and Other Small Objects*. Jerusalem, Israel Museum.

Stern, E.M. (1999) Roman glassblowing in a cultural context. *American Journal of Archaeology* 103.3, 441–484. doi: 10.2307/506970.

Vallet, F. (1988) A propos des tombes à épées d'apparat de La Rue-Saint-Pierre (Oise) et d'Arcy-Sainte-Restitue (Aisne). *Revue archéologique de Picardie* 3.1, 45–55. doi: 10.3406/pica.1988.1530.

Chapter 6

Identifying episodes of recycling in the archaeological record

Jonathan R. Wood

Incomplete data

According to a recent meme, there are two types of people in this world: 1) Those who can extrapolate from incomplete data.

This is obviously flippant. Nonetheless, archaeology has no choice but to deal with incomplete data, and it could be argued that it became an academic discipline because of it. Archaeological science attempts to increase the amount of archaeological data. Analysing the chemical compositions of artefacts, for example, can provide information about objects that were invisible to the people who made them.

More data, however, can lead to competition between data types. For example, A. Bernard Knapp (2012, 14–25) provides an illustrative example of a copper axe/adze (KM 457) and a piece of copper ore (KM 633) found at Kissonerga-Mosphilia (Late Chalcolithic/Pre-Bronze Age 1), a site which provided crucial data for the earliest stages of indigenous metalworking and casting activities in Cyprus. Lead isotopic analyses (LIA) conducted on the ore were consistent with Cypriot sources, but the copper axe was not (Gale 1991; Fig. 6.1). The axe's LIA signature does not overlap with Cypriot ores. It falls within hand-drawn and constructed ellipses but not within the kernel density estimates (KDE) that aim to represent Cyprus' LIA ore field; in fact, it lies closer to the ores of the Troad in northwest Anatolia (Pernicka *et al.* 1984, 557–558; Wagner *et al.* 1985; Wagner *et al.* 1986; Gale 1991). Three years after the initial study, Zofia Stos-Gale and Noel Gale (1994) published a LIA data point for ore from Akoursos in the west of Cyprus (an area with known deposits of copper and malachite), just inland from Kissonerga, which plots midway between KM 457 and the Cypriot ore field. This new information was used to support that copper ores in Cyprus could have been exploited outside the Troodos belt and that the Kissonerga axe (KM 457)

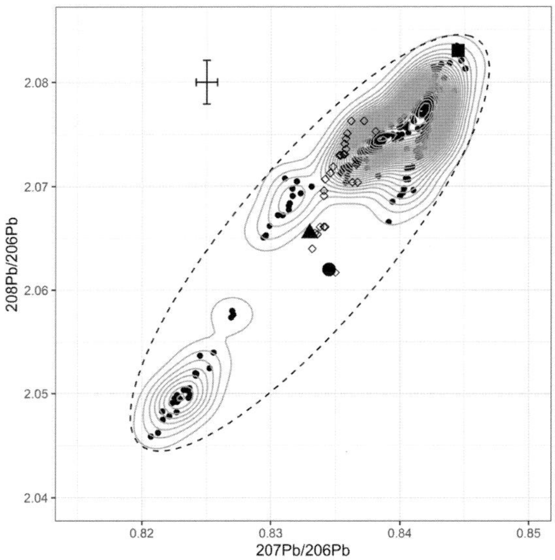

Fig. 6.1 LIA plot showing a copper axe/adze (KM 457) (large filled circle) and a piece of copper ore (KM 633) (large filled square) found at the Kissonerga-Mosphilia (Late Chalcolithic/Pre-Bronze Age 1), Cypriot ores (small filled circles) (OXALID 2022), ore from Akoursos (large filled triangle) (Stos-Gale and Gale 1994, 107) and ores from north western Anatolia (open diamonds) (Wagner et al. *1985). The contours represent the kernel density estimate for the Cypriot ores. The ellipse drawn around the Cypriot ore dataset includes all LIA ore data, but it also includes areas without ores, thereby losing its discriminatory power. Note that ^{204}Pb values were not available for the Akoursos ore, the copper axe/adze (KM 457) or the copper ore (KM 633).*

was made from local ores, thereby negating the need for a foreign ore source requiring foreign contacts.

Axe/Adze KM 457, however, is considered typologically distinct within Cypriot metalwork: it has a significantly broader butt than the few known early axes from Cyprus, which suggests that it may not be Cypriot (Dikaios in Gjerstad 1972, fig. 84; Stewart in Gjerstad 1972, 249). In fact, in terms of size, it falls comfortably within the range of those found in the Aegean (Croft *et al.* 1998), thereby supporting a north Anatolian origin.

This example highlights that different types of data can be in conflict. It also shows that the same type of data, in this case LIA measurements, can be interpreted differently, particularly for provenance investigations. Furthermore, ruling one location out does not mean necessarily that the other should be ruled in, particularly when there are differences in the amounts of LIA sampling that make the number of viable alternatives difficult to determine without archaeological evidence. Disagreements over the provenance of KM 457 have yet to be resolved.

Hierarchy among data can also manifest itself in the methods of measurement and the analytical procedures applied to interpret the data they generate. For instance, Noel Gale *et al.* (2003) dealt with discrepancies in LIA measurements of Bulgarian ores, between those measured by the Oxford group and those by another laboratory (Amov *et al.* 1979), by only including data which clustered with their own. This was classified as 'safe practice' (Gale *et al.* 2003, 155). However, neither Amov's laboratory nor the Oxford group found the extreme spread of data that was found in a third dataset by the Mainz group (Pernicka *et al.* 1993): that for Rudna Glava, attributed to a wide range in the U/Pb ratio of these ores. As the Oxford and Mainz groups took samples from different parts of the same shaft in the mine, the differences in values suggested that there were large differences in LIA measurements *within one mine*.

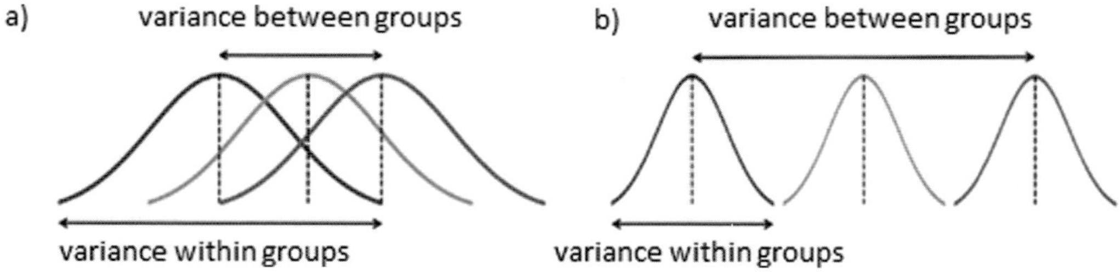

Fig. 6.2 Distributions of a continuous trait: a) the variance within groups is greater than the variance between groups; b) the variance between groups is greater than the variance within groups.

The issue here is not that different research groups using different instruments can produce conflicting results (unfortunately, this is common), but that these differences can threaten the provenance postulate (that is, variation within a source should not be greater than the variation between sources; Fig. 6.2) and, therefore, the application of LIA to trace objects to their geological sources. As there was less variation in the Oxford group's measurements than the Mainz group's, it was considered that the Oxford group's samples might have been contaminated with exogenous lead prior to arriving at the laboratory (Gale *et al.* 2003, 156).

Nonetheless, the differences in the results obtained for the Bulgarian ores highlight that placing too much emphasis on one type of data can be problematic. This is particularly evident in the research conducted on the ritual bronzes of the Chinese Shang dynasty (*c.* 1500–1046 BC), which identified sources in Africa as a probable region for their provenance, by matching highly radiogenic LIA signatures (Sun *et al.* 2016; Sun *et al.* 2018). This radical proposition has been strongly refuted by other researchers (Liu *et al.* 2018a; 2018b) applying archaeological and geological data (and by recognising that the distance travelled was probably prohibitive for this Bronze Age society). Furthermore, a recent multivariate approach (Wood and Liu 2022) that analysed both lead isotopes and elemental data from these ritual bronzes has shown that interpretations emerging from LIA and elemental data are in conflict: whereas lead isotope analyses can suggest various origins around and beyond China, elemental analyses suggest that the same ore deposits were exploited throughout the Chinese Bronze Age – potentially ore deposits that lie within the heartland of the Shang and Zhou dynasties. In effect, if there is systematic zoning of radiogenic and common lead within a deposit (Cannon *et al.* 1963), any change in isotopic signature could simply mark the expansion of mining in the same deposit rather than signalling changes in geographical location (Wood and Liu 2022); that is, lead isotopes can lose their discriminatory power to provenance from ore to object.

Recycling

Matching chemical signatures can be further complicated by recycling. Until recently, recycling was regarded as being not particularly significant in prehistory. For example, the practice of burying silver objects with their owners, thereby removing silver from circulation, was believed to have sustained a demand for new silver during the Aegean Bronze Age (Gale and Stos-Gale 1981). Furthermore, isotopic matches between oxhide ingots recovered around the Mediterranean and copper ore sources in Cyprus were considered sufficient to negate the possibility that similar signatures could emerge by mixing metals from different sources (Gale 2001). Recycling was also the first hypothesis to be rejected in favour of an assumed chemical compositional profile of an unprovenanced geological source to explain the presence of cobalt-copper glass in Egypt and Mycenae (Smirnou and Rehren 2013). As Mark Pollard *et al.* (2018, 38) highlight, many studies 'have either ignored completely the potential complexities arising from the mixing or recycling of metal from different sources, or have acknowledged that such practices would invalidate the simple hypothesis of provenance, but have then largely dismissed the problem.'

Recycling is difficult to identify when mixing is involved. However, examination of continuous data, such as chemical and isotopic data, can be used to identify mixed signatures associated with episodes of recycling. Nonetheless, despite being common in the archaeological literature, care must be taken with bivariate plots of elements and compounds used to identify mixing lines, as their application can lead to erroneous interpretations (Wood 2022).

To demonstrate this point, the bivariate plot shown in Figure 6.3 plots iron oxide against silica for glazes from Parthian and Sasanian glazed pottery (Wood and Hsu 2020). As these glazes are from many different archaeological sites, they are unlikely to have derived from the same melting pot. Nevertheless, a linear relationship can emerge (Fig. 6.3). Here, the three graphs plot *exactly* the same data (FeOwt% vs SiO_2wt%): the first plot closes the data to 1 (or 100%) for the following variables (SiO_2, Al_2O_3, FeO, MgO, CaO, K_2O and Na_2O); the second plot closes the data around the first four oxides (SiO_2, Al_2O_3, FeO and MgO); and the third plot closes the data around the two plotted variables (SiO_2 and FeO). The plots clearly show that closing a sub-composition does not necessarily produce similar or compatible patterns, even if some of the components are irrelevant to the variables being studied. This occurs because of a geometrical artefact of compositional data: the relationship between variables is not a real relation but one which is compelled, *i.e.* as the proportion of one component increases, the proportions of other components *must* decrease, which in the extreme case of a two-component system would always result in a correlation coefficient of -1. Although this 'spurious correlation' (Chayes 1949) may seem evident to many researchers, it is perhaps worth reiterating that it means that any regression lines drawn to represent the interaction between the two components (in this case SiO_2 and FeO) cannot be trusted, *i.e.* such an induced correlation may conceal true relationships among variables. Furthermore, the plots highlight that absolute compositional percentages are difficult to compare when different numbers of components are analysed. This issue has been approached by applying logratio analysis (Wood and Greenacre 2020).

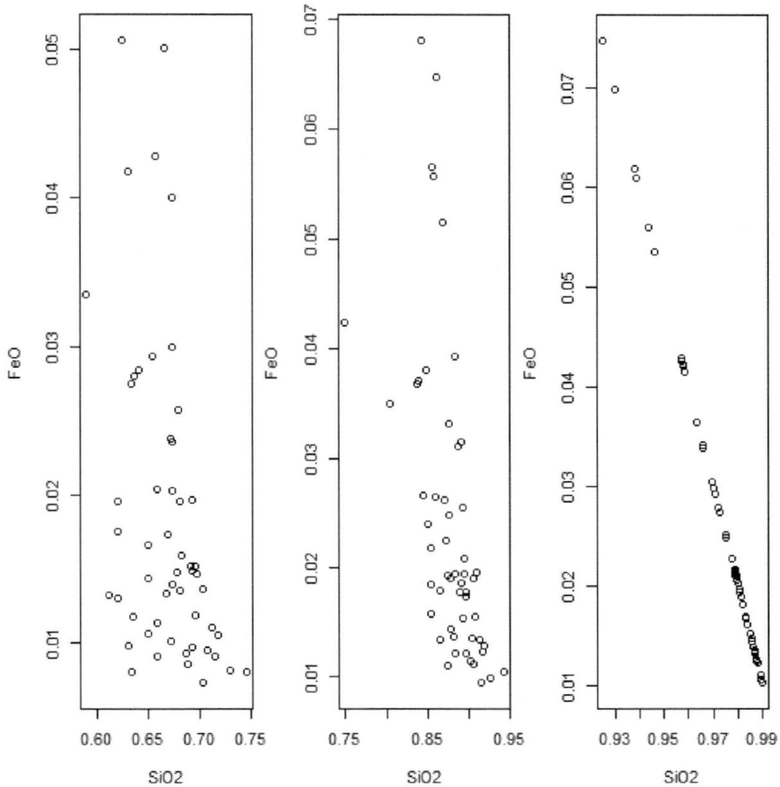

Fig. 6.3 Plots of glaze legacy data normalised to 1 (or 100%): closed around seven oxides (left), four oxides (middle) and two oxides (right). Note the change in scales and the emergence of linearity in the data as the number of elements used to close the data decreases. Refer to Wood and Hsu 2020 for details regarding the legacy data.

In addition to issues surrounding mathematical operations of closure or normalisation, linearity can emerge erroneously from plotting mathematically inconsistent combinations of compositional variables. Figure 6.4 shows data plotted by Ian Freestone (2015) for one of the few large sets of data from the Roman period that include analyses from a single archaeological assemblage of both antimony- and manganese-decolourised glasses, together with green-blue glasses (Silvestri 2008; Silvestri *et al.* 2008). This assemblage is the glass in the cargo of the lulia Felix, a ship that sank in the Adriatic in the 2nd or 3rd century AD.

On first inspection, the plot of $MnO/(Sb_2O_3+MnO)$ vs. Al_2O_3 produces what Freestone (2015) considered to be a mixing line between batches of low-alumina-antimony decolourised glass and high-alumina-manganese decolourised glass (Fig. 6.4). However, there is no reason for the percentages of decolourisers in glass to add up to a unique value, which makes limiting values difficult to interpret: that is, if MnO approaches zero, $MnO/(Sb_2O_3+MnO)$ approaches zero, regardless of the amount of Sb_2O_3; similarly,

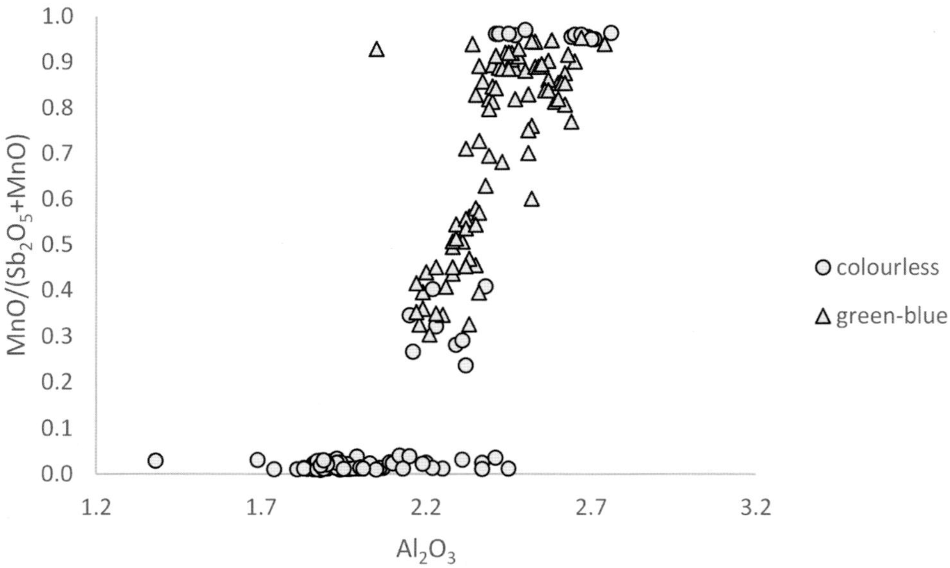

Fig. 6.4 Proportions of manganese and antimony in the glass of the Iulia Felix versus the alumina content. The data was considered to show a mixing line between the low-alumina antimony-decolorised glass and the high-alumina manganese-decolorised glass (Freestone 2015). Data from Silvestri 2008 and Silvestri et al. *2008.*

when Sb_2O_3 approaches zero, $MnO/(Sb_2O_3+MnO)$ will always approach 1, regardless of the amount of MnO. Furthermore, when the denominator is zero, the limiting value is undefined. In effect, although mixing may have occurred, forced linearity can affect archaeological interpretations as it could imply, perhaps incorrectly, the mixing of two components (or batches) rather than simply non-structured variation.

To address issues such as these, compositional data can be combined with lead isotope analytical data as both can reflect episodes of recycling but are *not* interdependent. For example, silver is often geologically associated with low concentrations of gold which, due to its chemical inertness, can survive silver smelting and refining operations. Moreover, silver is often geologically linked with lead, and lead is often used in silver extraction and cupellation. The isotopic signatures of the lead and the concentrations of gold associated with silver can therefore be used to provenance silver (Wood *et al.* 2019).

This method of combining compositional data and LIA data has been used to investigate silver from Iron Age hoards in the southern Levant (Fig. 6.5) by applying the following strategy (Wood *et al.* 2017; Wood *et al.* 2019): (1) LIA values are used to calculate the Pb crustal age (also known as the model age) of the ore, measured in millions of years (Ma), from which the silver (or, more correctly, the lead associated with the silver) derived; this age is then (2) plotted against the levels of gold found in

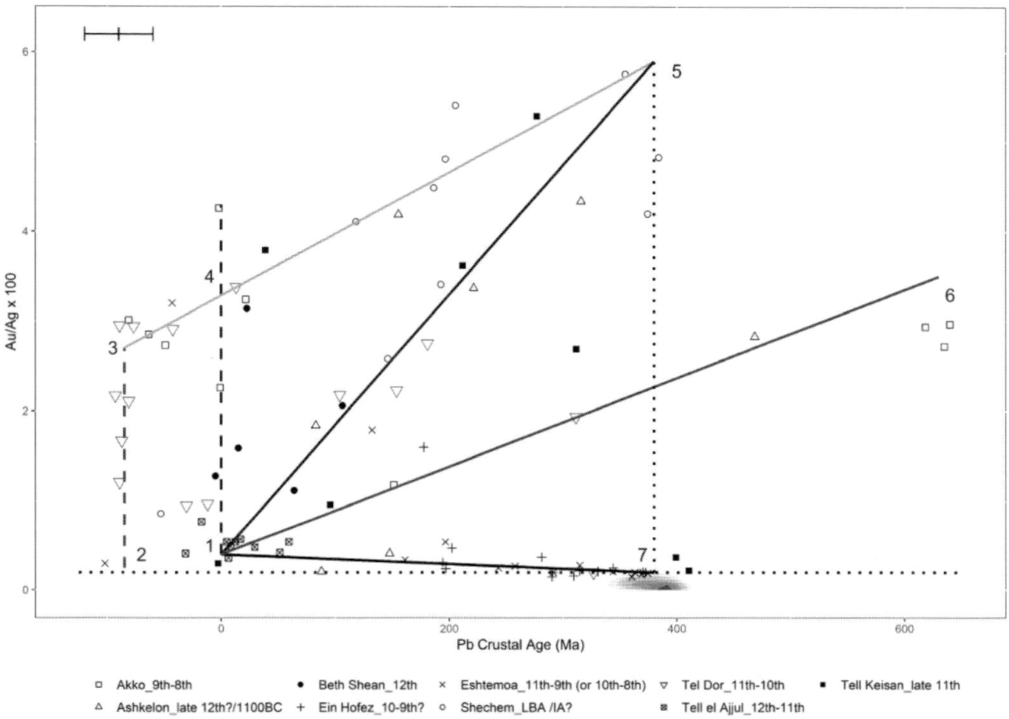

Fig. 6.5 Au/Ag × 100 versus Pb crustal age (Ma) for the Late Bronze Age/Early Iron Age silver hoards of the southern Levant. The density map at around 380 Ma is derived from compositional data (measured by ED-XRF) (OXALID 2022) and lead isotope data for Bronze Age silver found in Iberia (Bartelheim et al. *2012; Murillo-Barroso 2013; Comendador Rey* et al. *2014, 285–308; Murillo-Barroso* et al. *2015). For further details refer to Wood* et al. *2019.*

silver, given that gold is a useful geological indicator for a silver source as it is likely to survive silver processing. This approach allows two types of data to be plotted together, providing two parameters which would encourage any variation due to the mixing of components to form mixing lines. In effect, since time (Pb crustal age) and concentration (Au/Ag) are linear variables, linear mixing lines should emerge when two components are mixed. In effect, if it is assumed that endmembers are non-recycled silver (which, in some cases, can be related to archaeologically attested locations that produced silver), the relatively linear lines represent mixtures of these components.

Recycling and silver isotopes

Recently, archaeological science has approached silver coinage through the lens of geological provenance (*e.g.*, Birch *et al.* 2020; Albarède *et al.* 2021; Milot *et al.* 2022; Vaxevanopoulos *et al.* 2022). However, as with the hacksilver from the southern Levant in Figure 6.5, it needs to be recognised that *any* object which could be thrown back into a melting pot might have been recycled before it entered the archaeological record. Furthermore, an archaeological artefact made from recyclable material, such as glass or metal, may neither be the original object nor resemble it, with the consequence that a successful ancient recycling economy can become invisible when recycling is synonymous with re-melting rather than reuse (Freestone 2015; Ponting and Levene 2015).

Silver isotopes have been proposed as possible geochemical tracers of silver for silver coinage (*e.g.*, Desaulty *et al.* 2011, Albarède *et al.* 2016; Vaxevanopoulos *et al.* 2022) in part because lead used for refining silver does not necessarily derive from the same location as the source of the silver (that is, LIA signatures are likely to reflect the source of the lead rather than the source of silver). Unfortunately, Ag isotopic variation appears to be controlled by ore genesis rather than geology (Arribas *et al.* 2020), making them of marginal interest for provenance studies (Milot *et al.* 2022). Despite this limitation, some research groups have continued to pursue silver isotopes for provenance investigations (Albarède *et al.* 2021; Eshel *et al.* 2022; Vaxevanopoulos *et al.* 2022). Galena ores from Iberia, for example, have been proposed by Jean Milot *et al.* (2022) as the source of silver for Roman *denarii* of different origins and ages of minting, based on the narrow range of ε^{109}Ag (that is, the deviation per 10,000 of the ^{109}Ag isotope of a given sample relative to the U.S. National Institute of Standards and Technology [NIST] 978a Ag standard), compared to the wider range of ε^{109}Ag in silver-bearing base metal (Pb, Zn, and Cu) ores. Concentrations of antimony, bismuth, and arsenic in the silver coins, which have been attributed to pre-existing galena acting as a trap for fluids containing these elements in argentiferous ores (Milot *et al.* 2022), have been considered to further support this provenance.

Despite the application of two types of data (isotopic and elemental), this geological mechanism perhaps becomes moot when archaeological/historical evidence and recycling is taken into consideration. First, although it is well-known that Iberia was a major source of silver for the Phoenicians, Carthaginians and Romans, silver was derived from deposits of silver-bearing jarosite ore (especially around the mining region of Riotinto in southwest Iberia) rather than solely argentiferous lead ores, such as galena (*e.g.*, Rothenberg and Blanco-Freijeiro 1981; Hunt Ortiz 2003; Anguilano *et al.* 2009; Murillo-Barroso *et al.* 2015; Wood and Montero-Ruiz 2019). Roman lead ingots stamped with the Cartagena mark (Salkield 1982, 137–147; Salkield 1987, 15), some 400 km away in southeast Iberia, recovered at Riotinto and production slag (Anguilano 2012) exhibiting a potential influence of a Cartagena LIA signature, support that lead was added to these lead-poor jarosite ores to extract silver.

Second, silver was often debased with copper to make coinage (Butcher and Ponting 2015). This debasement process could have conceivably introduced antimony and arsenic into the silver if the copper derived from fahlore (that is, a solid solution of tetrahedrite $Cu_{12}Sb_4S_{13}$ and tennantite $Cu_{12}As_4S_{13}$; Pernicka *et al.* 2016). Moreover, concentrations of bismuth in silver coinage are perhaps better explained by recognising that the lead used in the cupellation process to refine silver can contain bismuth (L'Heritier *et al.* 2015).

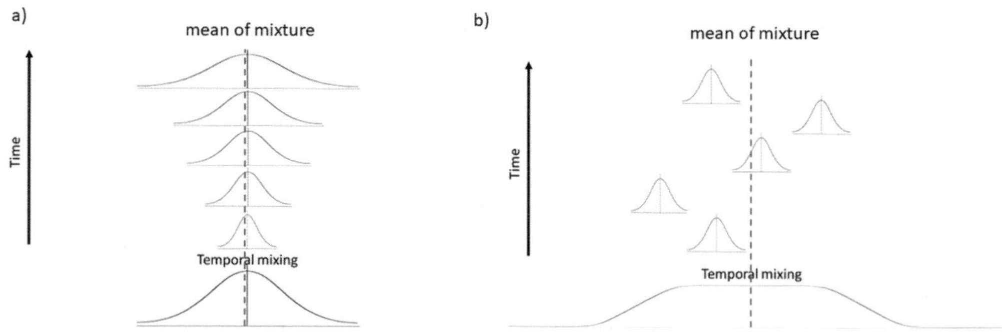

Fig. 6.6 Mixing increases variance: a) a stable mean and fluctuating variance (i.e., *the mean of the mixture remains constant); b) a fluctuating mean and stable variance* (i.e., *the means of the individual groups converge in the mixture). (Adapted from Perreault 2019, fig. 3.12).*

Third, and perhaps most significant, silver coinage was most likely recalled, reminted, and reissued (Butcher and Ponting 2015). Mixing of silver and/or silver coinage would have affected elemental and isotopic signatures. Figure 6.6 presents changes associated with mixing for any continuous variable (Perreault 2019) and highlights that the provenance postulate (Figs 6.2a and 6.2b) can be affected by mixing. Figure 6.6a shows that if silver and/or silver coins all derived from one silver source and were smelted and refined using the same raw materials over time, the mean would remain relatively constant after recycling, even if differences in processing (*e.g.*, between batches) caused fluctuations in the variance of the distributions. In fact, a stable mean but changing variance would result in a time-averaged distribution in which the central values are overrepresented. However, if silver coins from different origins were mixed together (perhaps during reminting and/or debasement), the mean values of the coins would likely fluctuate over time and the variance in the time-average assemblage would be much larger than it ever was at any single point in time (Fig. 6.6b): that is to say, even for a constant variance, the distribution of the mixture would be wider, flatter and converging to a uniform distribution. The mean of this mixture would lie between the means of the individual batches.

Essentially, recycling can explain the narrow range of silver isotope values with respect to the standard (that is, ε^{109}Ag) not only for Roman coins (Milot *et al.* 2022) but also for Hellenistic, medieval and modern silver coins and hacksilver (Desaulty *et al.* 2011; Desaulty and Albarede 2013; Albarède *et al.* 2016; Albarède *et al.* 2021; Milot *et al.* 2021; Eshel *et al.* 2022; Milot *et al.* 2022; Vaxevanopoulos *et al.* 2022) which all report ε^{109}Ag values that have narrower ranges than those reported for hypogene and supergene ores (Mathur *et al.* 2018; Arribas *et al.* 2020). This would cast doubt on the regions in Iberia identified for the provenance of the silver used for Roman coinage (Milot *et al.* 2022), or at least the procedure used to arrive at this provenance.

In summary, refining, debasement, and recycling of silver coinage would potentially negate the need for a geochemical mechanism to explain the concentrations of antimony, arsenic and bismuth, and the narrow range in the silver isotopic signatures of coins compared to those found in argentiferous ore deposits. To some extent, the issues raised here regarding the interpretation of silver isotopes when recycling is involved are similar to those levelled against lead isotopes over 40 years ago.

Approaches to dealing with recycled objects

Recycling is now being addressed by not focusing exclusively on the provenancing of objects (Pollard *et al.* 2018) and through the development of theoretical frameworks which advocate the inclusion of the mutability of artefacts in the chaîne opératoire (Sainsbury *et al.* 2021). Nonetheless, before any discussion can be held on the motivating factors for and the implications of recycling, it is necessary to identify whether the artefacts, which provide evidence for such discussions, have been recycled or not.

There are two main ways in which archaeology can approach recycling: the first is to identify material that is out of place in an archaeological context which can be deduced to have been recycled; the second is by investigating the recycling event (such as those events suggested by the emergence of mixing lines in Figure 6.5), as episodes of recycling can result in the deliberate or accidental introduction, dilution, and loss of components compared with those in the original object. Both scenarios require that something is known about the original non-recycled material, such as its chemical composition (often referred to as an endmember when discussing recycling in terms of mixing) before it can be suggested that recycling must have occurred.

Out-of-context material

An example of an out-of-context material is that of Roman glass being recycled and reused to glaze Parthian pottery. First, the archaeological context supports that this process could have occurred: Parthian glazed pottery was recovered at the Roman military outpost of Ain Sinu in northern Mesopotamia (modern-day Iraq), which was contested territory for both the Parthian and Sasanian empires and was occupied by the Romans between AD 197 and 364 (Oates and Oates 1959). Second, some of the Parthian glazes recovered at this site exhibit not only different chemical compositional signatures to other Parthian and Sasanian glazes (Fig. 6.7a–c) but also have similar compositions to Roman glass (Table 6.1); that is, since Roman glaze production used lead-based glazes (Hatcher *et al.* 1994; Walton and Tite 2010), this would indicate that Roman natron glass was recycled to glaze Parthian pottery.

The analyses of the glaze data are presented elsewhere (Wood and Hsu 2020; Wood and Greenacre 2021). However, it is perhaps the differences between Roman glass and the glazed pottery that are most relevant to recycling. The K_2O concentration is several times higher in the 'natron' glazes than for Roman glass. This could indicate that another source of potash entered the glaze systems, other than that associated with plant ash used for Mesopotamian glazed pottery, *i.e.*, a source unrelated to the initial

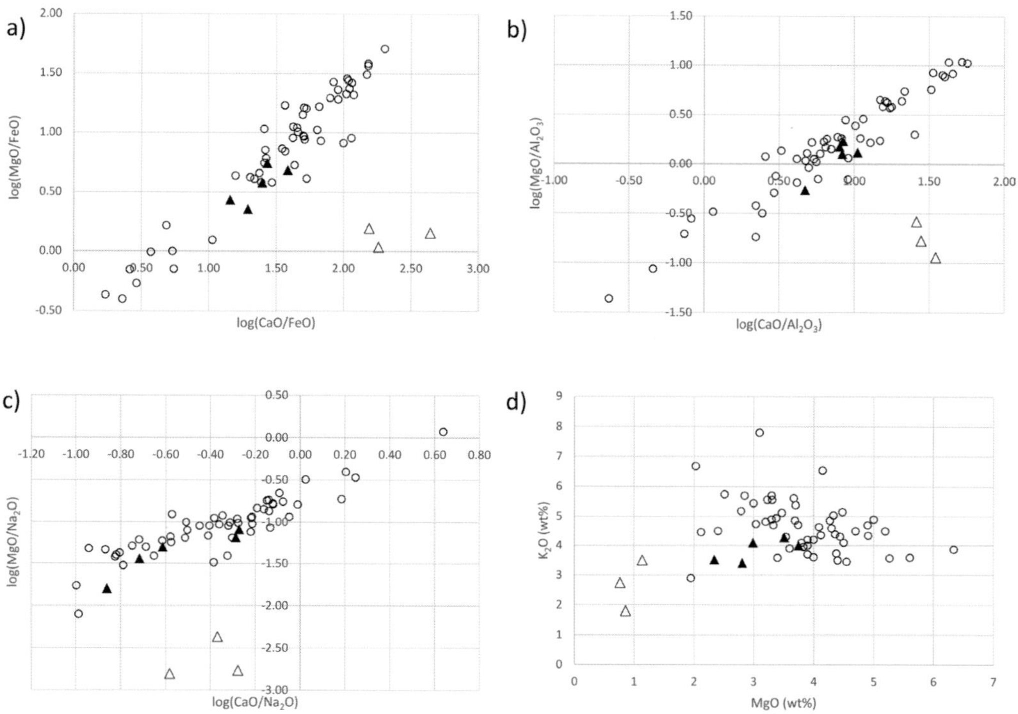

Fig. 6.7 (a–c) Logratio plots of the glaze compositions for the legacy data and the Ain Sinu glazed sherds. (d) is a bivariate plot of K_2O (potash) vs. MgO (magnesia). Open circles are from the legacy data. Triangles show the samples from Ain Sinu (Wood and Hsu 2020). Open triangles are those glazes from Ain Sinu which appear to have been fluxed using natron, i.e., Roman glass.

Table 6.1 Composition of typical Roman glass (Freestone 1991) with Sb_2O_5 values from Jackson and Paynter (2016) and the mean compositions of the three 'natron' samples which fall outside the legacy data bands and the remaining five samples from Ain Sinu (Fig. 6.7). n denotes the number of sherds. BDL denotes levels below the detection limit. Note the similarity between the Ainu Sinu 'natron' glazes and the typical composition of Roman glass for these oxides, apart from K_2O.

	Na_2O	MgO	SiO_2	K_2O	CaO	Sb_2O_5
Typical Roman glass (Freestone 1991)	16	0.5	68	0.5	8	0.54
Ain Sinu 'natron' glazes (n=3)	12.88	0.92	68.10	2.68	8.56	0.41
Ain Sinu (n=5)	11.93	3.08	64.84	3.86	6.97	BDL

composition of the glaze. Sarah Paynter (2009) suggests that differences between Mesopotamian glasses and glazes could be a consequence of contamination from fuel ash or vapour, especially hard-wood fuel. Since glazes have a higher surface area-to-volume ratio compared to glass heated in a crucible, ash or vaporised potassium compounds are more likely to affect the composition of a glaze by adhering to the molten surface. This would suggest that levels of K_2O in the glazes investigated (legacy and Ain Sinu glazes) could be a consequence of potassium compounds from the plant ash flux and the type of fuel used in Parthian and Sasanian kilns. This is supported by the invariance of K_2O with increasing concentrations of MgO on a bivariate plot for the legacy data and the majority of the Ain Sinu glazes (Fig. 6.7d). Essentially, the contamination argument outlined above could suggest that the three 'natron' samples (open triangles in Fig. 6.7) were produced in the same tradition as the other glazes, but that the starting composition of the glass was different to the other glazes from Ain Sinu and to those of the legacy data.

To some extent this scenario feeds into the theoretical approaches on mutability (Sainsbury *et al.* 2021), in that changes in function can be viewed through variation in chemical markers that occur during recycling (in this case, increases in K_2O with the change from glass to glaze) in a similar fashion to how volatile compounds such as arsenic compounds found in copper alloys (McKerrell and Tylecote 1972; Bray and Pollard 2012) and in cobalt-blue glass (Wood and Hsu 2019) can be lost on remelting. In effect, the three samples from Ain Sinu started out with lower concentrations of K_2O (because they are Roman glass, fluxed using natron) but were contaminated with K_2O because they were produced in the same tradition as the other glazes.

The presence of Roman glass reapplied as a glaze on Parthian pottery is a good example of a material which is out of context and provides convincing evidence that glass was being recycled at Ain Sinu. In fact, if Parthian potters had not used Roman glass, it would have been much more difficult to identify that recycling had occurred. As only eight samples from Ain Sinu were offered by the British Museum for testing, it could be argued that the data generated from these few samples should be used to support a larger study. However, it is perhaps here that archaeological science needs to resist attempts to create a series of compositional typologies for materials from various chronologies and regions; that is, instead of searching for artefacts that support the hypothesis that 'all swans are white' (Popper 2002, 82–83) it may be better to look for the 'black swan' and recognise that ancient recycling will disturb any attempts at neat categorisation. In effect, archaeological science might be better utilised to investigate the repercussions of exceptions. For instance, extending recycling from Roman glass to indigenous glass production could provide an additional reason to weathering of glass as to the relatively low numbers of Parthian and Sasanian glass artefacts in the archaeological record (that is, each glass vessel could have been recycled to glaze two or three pots of equivalent size). Furthermore, evidence of recycling indicates that glass and glaze were regarded as the same material in Mesopotamian production and that potters made their own frit specifically to glaze pottery, with recycling potentially reflecting interruptions in supply chains from, as yet undiscovered, glass production sites in Mesopotamia.

Recycling with mixing

Variation in compositional data of artefacts can be interrogated using histograms (*e.g.*, Pollard *et al.* 2019; Wood and Hsu 2019), with the caveat that samples must derive from many different objects. This condition may sound trivial; however, differences in compositional signatures from, say, sherds of glass or glass beads found in the same context are unlikely to represent variation related to provenance or recycling but to represent variation in processing within a batch. Furthermore, the underlying assumptions behind distributions can affect the identification of recycling, as mixing affects distributions by increasing variance (Perreault 2019). This increase is potentially irrespective of whether mixing was a consequence of an ancient melting pot (Fig. 6.6) or whether archaeologists combine artefacts from several contexts into one mixed assemblage after excavation. If several archaeological contexts deposited at different points in time are analysed together, the variance within each chronological context will collapse into the variance that exists between chronological contexts. In fact, a similar increase in variance would happen if the same process occurred when different locations (perhaps on the same site) were combined, that is, different areas with different means and/or variance from the same chronological setting. Essentially, an archaeologist measuring and recording a continuous trait, such as the diameter, weight, or chemical composition of artefacts etc., may combine different contexts, probably inadvertently, in the analysis. The greater the between-contexts variance, the more inflated the variance would be when assemblages are mixed together (Fig. 6.6b).

A punctuated shift in the mean or a gap in the sequence of mixed contexts, however, would lead to a bimodal distribution. Figures 6.8a and 6.8b show slow change punctuated with a period of rapid change. For compositional data, this means that if it can be assumed that punctuation is not a consequence of missing data (*i.e.*, the absence of part of the sequence), a bimodal distribution for compositional data could suggest a sudden change in recipe.

The following histograms (Fig. 6.9) plot cobalt-blue glass from New Kingdom Egypt (Wood and Hsu 2019), showing a very similar pattern to those in Figure 6.8. The upper plot shows 30 compositional analyses from Abe *et al.* (2012): 26 different 18th-Dynasty objects from museum collections in Japan, provenanced to Malkata and Amarna, and four 18th-Dynasty objects recovered at the site of Dashur. The lower figure plots 20 compositional analyses from different objects from 19th–

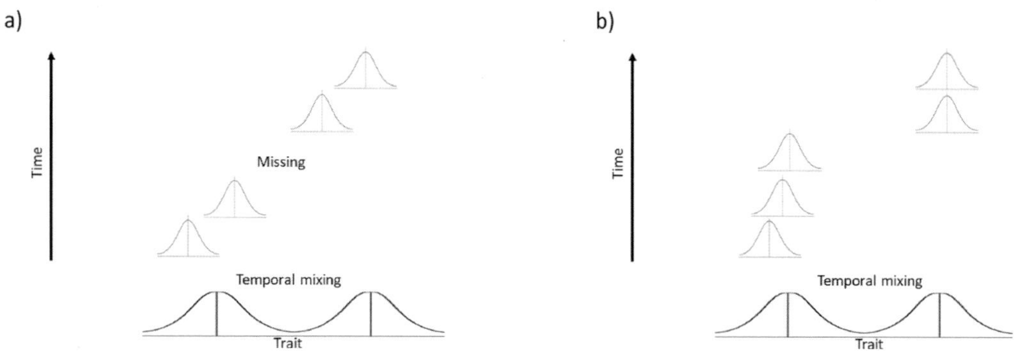

Fig. 6.8 Mixing increases variance in archaeological assemblages: a) Noncontinuous time averaging; b) Slow change with punctuated period of rapid change (adapted from Perreault 2019, fig. 3.12).

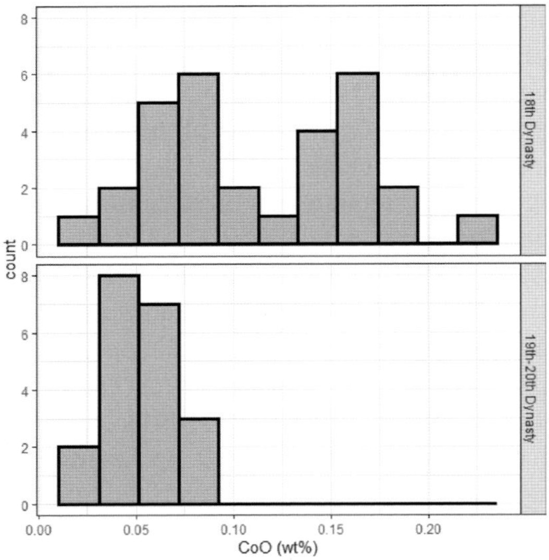

Fig. 6.9 Frequency histograms of the cobalt oxide (CoO %) in cobalt-blue glass from Egypt in the 18th and 19th–20th Dynasties. Data from Abe et al. *(2012). The upper plot exhibits a bimodal distribution with the higher peak having about twice the concentration of CoO as the lower peak from the 18th-Dynasty objects. The 19th–20th-Dynasty cobalt glass has similar levels of CoO to the lower peak from the 18th-Dynasty cobalt glass.*

20th-Dynasty contexts at Dashur. Although differences in composition between the two periods have been attributed to the exploitation of a new cobalt source in the 19th Dynasty (Abe *et al.* 2012), it is proposed that glass recycling is perhaps a more parsimonious interpretation.

Figure 6.9 (upper) shows that the distribution of cobalt oxide concentration in the 18th-Dynasty glass appears to be bimodal, with the lower peak having about half the CoO level of the higher peak. The 19th–20th-Dynasty glass (Fig. 6.9 – lower) has a distribution which is similar to the lower peak of the 18th-Dynasty cobalt-blue glass. Assuming that the bimodality is not due to missing data (which seems a reasonable assumption for sites which have been thoroughly excavated), this could suggest that the original glass was the higher peak in the upper plot (*i.e.* the peak with a mean around 0.16wt%CoO), which was diluted in a 1:1 ratio with a base glass (*i.e.* glass without cobalt colourant) thereby diluting the amount of cobalt in the mixture to about 0.8wt%CoO. The 19th–20th-Dynasty glass objects were potentially made from this mixture (or perhaps from a 1:1 dilution of remelted glass from both distributions found in 18th-Dynasty contexts with a base glass). This dilution is also reflected in the concentrations of other oxides associated with the colourant, such as NiO and MnO (Wood and Hsu 2019). In effect, this bimodal distribution would be consistent with a recycling process which also included dilution. Although such dilutions were probably derived empirically, it is interesting to note that the cobalt level in the glass from 19th–20th-Dynasty Dashur (average CoO = 533ppm) was around the lower limit to maintain a rich, deep blue (Henderson 2000). This fits well with the narrative that glass was not produced in significant amounts after *c.* 1250 BC (Shortland 2016, 95), with glassmakers reducing the levels of cobalt through dilution with a base glass to the point where the colour was still effective. Recycling is also supported by broad distributions of PbO and Sb_2O_3, which indicate unintentional mixing due to the failure to separate all the white opaque glass (calcium antimonate) as well as yellow glass (lead antimonate) from cobalt-blue glass during recycling (Wood 2022).

Essentially, the compositional data suggests recycling and dilution (and its associated accidental introduction of contaminants) rather than a deliberate reduction in the amounts of cobalt added to new glass in this later period. It would also suggest that certain Egyptian and Mycenaean blue glasses, which have both copper and cobalt as colourants, are *not* deliberate mixtures to produce specific shades of blue or due to compositional variability in the geology of the colourant source (Smirnou and Rehren 2013) but rather the accidental mixing of different types of blue glass during recycling.

This mechanism further indicates that recycling became necessary in New Kingdom Egypt and that the change in composition of cobalt-blue glass was not because a new source of cobalt was exploited in later periods (Abe *et al.* 2012) but because cobalt sources were no longer available to provide the colour cobalt-blue. In effect, if a source was no longer available, but cobalt glass continued to be used, and a new source was not exploited, then recycling must have occurred. This would question conventional wisdom concerning the accepted sources of cobalt for the New Kingdom period, that is the Kharga and Dakhla Oases of Egypt's Western desert (*e.g.*, Kaczmarczyk 1986, 369–376; Shortland 2000; Rehren 2001), and raise the possibility that less local sources were used (*e.g.*, Lucas 1962; Dayton 1981; 1993; Stern and Schlick-Nolte 1994; Lee and Quirke 2000, 111; Wood and Hsu 2019) which would have been more susceptible to interruptions due to changes in the political landscape affecting supply chains.

Other compositional investigations such as those conducted by Varberg *et al.* (2015; 2016) demonstrate that glass of Mesopotamian origin reached Egypt. However, it is perhaps the archaeological evidence in Egypt that tips the balance regarding the provenance of the cobalt-blue colourant used for New Kingdom Egyptian glass. The Hall of the Annals at the Temple at Karnak, in Thebes, has a scene on one wall depicting Thutmosis III's spoils from his successful campaigns in the Levant and Syria (Fig. 6.10), including gold, silver and glass (Sherratt and Sherratt 1991), with glass referred to as 'stones of casting' in the Amarna letters. Three baskets are dark-blue, and their contents are depicted in two forms: one basket has five round objects, the other two are baskets of irregularly shaped lumps. The round objects appear human-made and are potentially dark-blue glass ingots (Shortland 2012, 140–145). Although one of the baskets with irregularly shaped lumps could be lapis lazuli, the second is denoted differently, which suggests that it is not lapis lazuli but is also dark-blue glass.

Concluding remarks

The main point made in this article is that recycling needs to be considered before any type of archaeological data is extrapolated to make interpretations about societies, economies, and trade, even though it can make interpretations less definitive. As Perreault (2019, 2) puts it, 'archaeologists often settle on an explanation on the sole basis it can be made consistent with their data, thereby ignoring the fact that there are a number of alternative explanations that are just as consistent with the data.' It

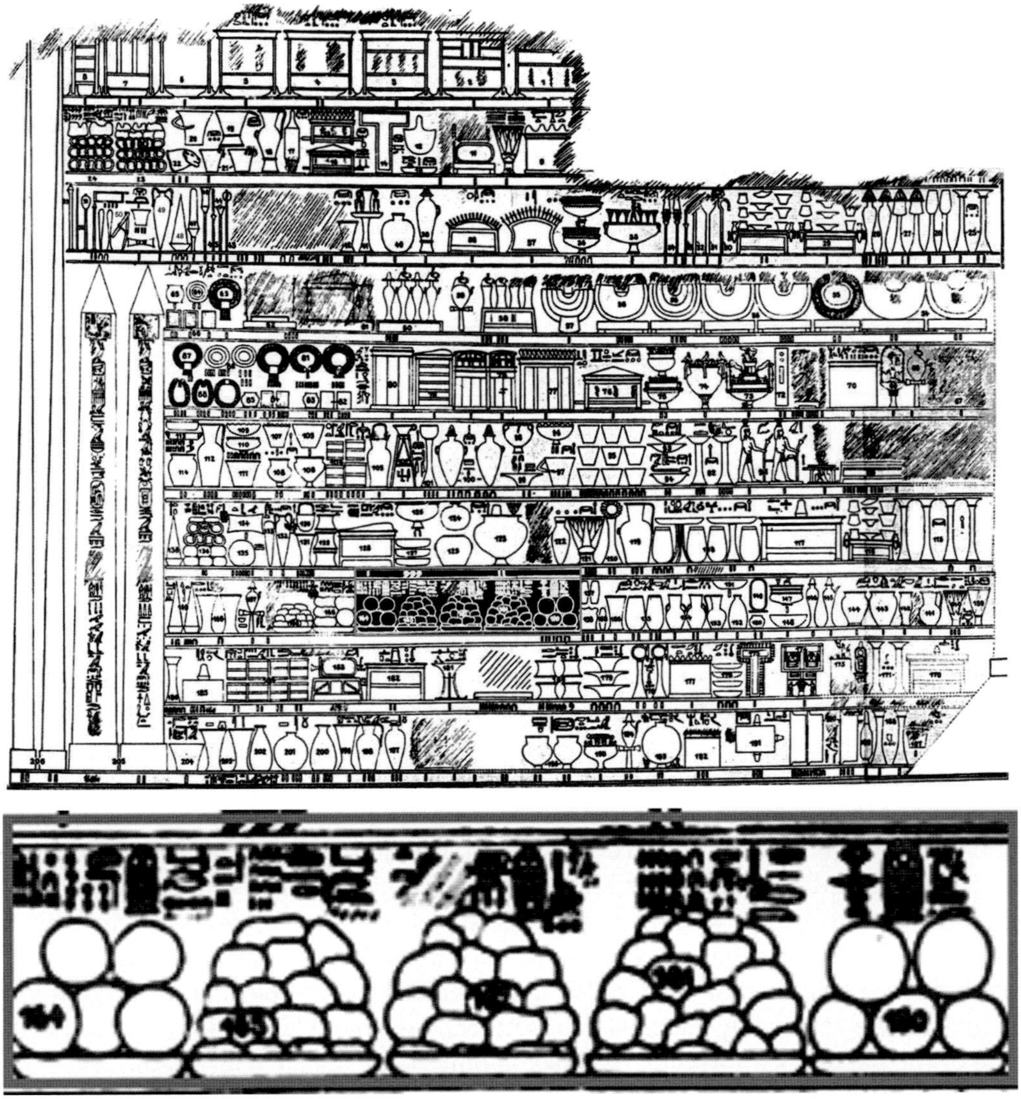

Fig. 6.10 (upper) Depiction of the Hall of the Annals scene at Karnak showing the tribute given by Thutmosis III to the temple. (lower) Close-up showing baskets of glass and precious stone. The scene reads right to left: basket 1 with the five round objects is at the far right with two baskets of irregularly shaped lumps to its left (adapted from a drawing in Wreszinski 1931). See also https://www.flickr.com/photos/manna4u/20629027214/in/photostream.

is clear that without considering recycling, the numbers of alternative explanations required for blurred elemental and isotopic signatures increases dramatically. However, this blurring cannot be attributed solely to variation in the signatures of geological sources; that is, although it may be possible to match the analytical signature of an artefact to a geological source (especially when the source exhibits high variation), this does not mean necessarily that the source was exploited to make that artefact. In effect, analytical signatures of artefacts should only be attributed to specific geological sources when there is archaeological and/or historical evidence to support that these sources were exploited at the time when the objects were made. Furthermore, this article highlights that placing too much emphasis on one type of data can lead to erroneous archaeological interpretations, which may remain in the academic literature long after the limitations of the techniques and analyses that produced the data are well known. Nonetheless, although there is no one-size-fits-all approach to investigate ancient recycling, there are methods that align with the data available that may provide insight well beyond elemental composition that can delimit and inform on raw material availability and supply chains over time, the process of manufacture and how these might relate to external economic and political circumstances. It is therefore imperative that recycled material can be identified with some certainty in the archaeological record.

Acknowledgements

This project has received funding from the European Research Council (ERC) under the European Union's Horizon 2020 research and innovation programme (grant agreement No. 835180).

Bibliography

Abe, Y., Harimoto, R., Kikugawa, T., Yazawa, K., Nishisaka, A., Kawai, N., Yoshimura, S. and Nakai, I. (2012) Transition in the use of cobalt-blue colorant in the New Kingdom of Egypt. *Journal of Archaeological Science* 39.6, 1793–1808.

Albarède, F., Blichert-Toft, J., Rivoal, M. and Telouk, P. (2016) A glimpse into the Roman finances of the Second Punic War through silver isotopes. *Geochemical Perspective Letters* 2, 127–137. https://doi.org/10.7185/geochemlet.1613

Albarède, F., Blichert-Toft, J., de Callataÿ, F., Davis, G., Debernardi, P., Gentelli, L., Gitler, H., Kemmers, F., Klein, S., Malod-Dognin, C., Milot, J., Télouk, P., Vaxevanopoulos, M. and Westner, K. (2021) From commodity to money: the rise of silver coinage around the Ancient Mediterranean (sixth–first centuries bce). *Archaeometry* 63.1, 142–155. https://doi.org/10.1111/arcm.12615.

Amov, B., Incheva-Stefanova, I.M. and Baldjieva, T. (1979) New data on the lead isotope composition of galenas from the stratabound polymetallic deposits in the western Stara Planina Mountain, Bulgaria. *Geokhimia* 4, 535–554.

Anguilano, L. (2012) *Roman lead and silver smelting at Rio Tinto: the case study of Corta Lago.* Unpublished thesis, University College London.

Anguilano, L., Rehren, T., Muller, W. and Rothenberg, B. (2009) Roman jarosite exploitation at Rio Tinto (Spain). In A. Giumlia-Mair and A. Hauptmann (eds) *Archaeometry in Europe II 2007*, 21–29. Milano, Associazione Italiana di Metallurgia.

Arribas, A., Mathur, R., Megaw, P. and Arribas, I. (2020) The isotopic composition of silver in ore minerals. *Geochemistry, Geophysics, Geosystems* 21.8, e2020GC009097. https://doi.org/10.1029/2020GC009097

Bartelheim, M., Contreras Cortes, F., Moreno Onorato, A., Murillo-Barroso, M. and Pernicka, E.

(2012) The silver production of the South Iberian El Agar culture; A first look at production and distribution. *Trabajos de Prehistoria* 69, 293–309.
Birch, T., Westner, K.J., Kemmers, F., Klein, S., Höfer, H.E. and Seitz, H.-M. (2020) Retracing Magna Graecia's silver: coupling lead isotopes with a multi-standard trace element procedure. *Archaeometry* 62.5, 81–108.
Butcher, K. and Ponting, M. (2015) *The Metallurgy of Roman Silver Coinage: From the Reform of Nero to the Reform of Trajan*. Cambridge, Cambridge University Press.
Bray, P.J. and Pollard, A.M. (2012) A new interpretative approach to the chemistry of copper-alloy objects: source, recycling and technology. *Antiquity* 86, 853–867.
Cannon, R.S.J., Pierce, A.P. and Delevaux, M.H. (1963) Lead isotope variation with growth zoning in a galena crystal. *Science* 142, 574–576.
Chayes, F. (1949) On correlation in petrography. *The Journal of Geology* 57, 239–254.
Comendador Rey, B., Millos, J. and Alvarez-Iglesias, P. (2014) Provenance of the prehistoric silver set of Antas de Ulla North-Western Iberia, using lead stable isotope ratios. In H. Meller, R. Risch and E. Pernicka (eds) *Metals of Power—Early Gold and Silver: 6th Archaeological Conference of Central Germany, October 17-19, 2013 in Halle (Saale)*, 285–308. Halle (Saale), Landesamt für Denkmalpflege und Archäologie Sachsen-Anhalt, Landesmuseum für Vorgeschichte.
Croft, P., Peltenburg, E. and Tite, M. (1998) Other artefacts. In E. Peltenburg, D. Bolger and E. Baxevani (eds) *Excavations at Kissonerga-Mosphilia 1979-1992, Volume II.1A. Lemba Archaeological Project, Studies in Mediterranean Archaeology Vol: LXX:2*, 188–201. Jonsered, Paul Åströms Förlag.
Dayton, J.E. (1981) Geological evidence for the discovery of cobalt blue glass in Mycenaean times as a by-product of silver smelting in the Schneeberg area of the Bohemian Erzgebirg. *Revue d'Archaeometrie* 1, 57–61.
Dayton, J.E. (1993) *The Discovery of Glass: Experiments in the Smelting of Rich, Dry Silver Ores, and the Reproduction of Bronze Age-Type Cobalt Blue Glass as a Slag*. Cambridge, Harvard University Press.
Desaulty, A.-M. and Albarede, F. (2013) Copper, lead, and silver isotopes solve a major economic conundrum of Tudor and early Stuart Europe. *Geology* 41.2, 135–138.
Desaulty, A.-M., Telouk, P., Albalat, E. and Albarède, F. (2011) Isotopic Ag–Cu–Pb record of silver circulation through 16th–18th century Spain. *Proceedings of the National Academy of Sciences* 108.22, 9002–9007. https://doi.org/10.1073/pnas.1018210108
Eshel, T., Tirosh, O., Yahalom-Mack, N., Gilboa, A. and Erel Y. (2022) Silver isotopes in silver suggest Phoenician innovation in metal production. *Applied Sciences* 12.2, 741. https://doi.org/10.3390/app12020741
Freestone, I.C. (1991) Looking into glass. In S. Bowman (ed.) *Science and the Past*, 37–56. Toronto, University of Toronto Press.
Freestone, I.C. (2015) The recycling and reuse of Roman glass: analytical approaches. *Journal of Glass Studies* 57, 29–40.
Gale, N.H. (1991) Metals and metallurgy in the Chalcolithic period. *Bulletin of the American Schools of Oriental Research* 282/283, 37–62.
Gale, N.H. (2001) Archaeology, science-based archaeology and the Mediterranean Bronze Age metals trade: a contribution to the debate. *European Journal of Archaeology* 4, 113–130.
Gale, N.H. and Stos-Gale, Z.A. (1981) Lead and silver in the Ancient Aegean. *Scientific American* 244, 176–193.
Gale, N.H., Stos-Gale, Z., Raduncheva, I., Panayotov, I., Ivanov, I., Lilov, P. and Todorove, T. (2003) Early metallurgy in Bulgaria. In P. Craddock and J. Lang (eds) *Mining and Metal Production through the Ages*, 102–168. London, British Museum Press.
Gjerstad, E. (1972) *The Swedish Cyprus Expedition IV.1A*. Lund, Swedish Cyprus Expedition.
Hatcher, H., Kaczmarczyk, A., Scherer, A. and Symonds, R.P. (1994) Chemical classification and provenance of some Roman glazed ceramics. *American Journal of Archaeology* 98, 431–456.
Henderson, J. (2000) *The Science and Archaeology of Materials: An Investigation of Inorganic Materials*. London/New York, Routledge.

Hunt Ortiz, M.A. (2003) *Prehistoric Mining and Metallurgy in South West Iberian Peninsula.* Oxford, British Archaeological Reports.

Jackson, C.M. and Paynter, S. (2016) A great big melting pot: Exploring patterns of glass supply, consumption and recycling in Roman Coppergate, York. *Archaeometry* 58.1, 68–95.

Kaczmarczyk, A. (1986) The source of cobalt in ancient Egyptian pigments. In J.S. Olin and M.J. Blackman (eds) *Proceedings of the 24th International Symposium on Archaeology*, 369–376. Washington D.C., Smithsonian Institution Press.

Knapp, B.A. (2012) Metallurgical production and trade on Bronze Age Cyprus: views and variations. In V. Kassianidou and G. Papasavvas (eds) *Eastern Mediterranean Metallurgy and Metalwork in the Second Millennium*, 14–25. Oxford, Oxbow Books.

Lee, L. and Quirke, S. (2000) Painting materials. In P.T. Nicholson and I. Shaw (eds) *Ancient Egyptian Materials and Technology*, 108–111. Cambridge, Cambridge University Press.

L'Heritier, M., Baron, S., Cassayre, L. and Tereygeol, F. (2015) Bismuth behaviour during ancient processes of silver-lead production. *Journal of Archaeological Science* 57, 56–68.

Liu, S., Chen, K., Rehren, T., Mei, J., Chen, J., Liu, Y. and Killick, D. (2018a) Did China import metals from Africa in the bronze age? *Archaeometry* 60.1, 105–117.

Liu, S., Chen, K., Rehren, T., Mei, J., Liu, Y. and Killick, D. (2018b) Lead isotope and metal source of Shang bronzes: a response to Sun et al.'s comments. *Archaeometry* 60.5, 1040–1044.

Lucas, A. (1962) *Ancient Egyptian Materials and Industries,* 4th edition. London, Edward Arnold.

Mathur, R., Arribas, A., Megaw, P., Wilson, M., Stroup, S., Meyer-Arrivillaga, D. and Arribas, I. (2018) Fractionation of silver isotopes in native silver explained by redox reactions. *Geochimica et Cosmochima Acta* 224, 313–326.

McKerrell, H. and Tylecote, R.F. (1972) Working of copper-arsenic alloys in the Early Bronze Age and the effect on the determination of provenance. *Proceedings of the Prehistoric Society* 38, 209–218.

Milot, J., Malod-Dognin, C., Blichert-Toft, J., Télouk, P. and Albarède, F. (2021) Sampling and combined Pb and Ag isotopic analysis of ancient silver coins and ores. *Chemical Geology* 564, 120028. https://doi.org/10.1016/j.chemgeo.2020.120028.

Milot, J., Blichert-Toft, J., Ayarzagüena Sanz, M., Malod-Dognin, C., Télouk, P. and Albarède, F. (2022) Silver isotope and volatile trace element systematics in galena samples from the Iberian Peninsula and the quest for silver sources of Roman coinage. *Geology* 50.4, 422–426. https://doi.org/10.1130/G49690.1

Murillo-Barroso, M. (2013) *Producción y Consumo de Plata en la Península Ibérica. Un Análisis Comparativo entre la Sociedad Argárica y los Primeros Asentamientos Orientalizantes.* Unpublished thesis, Universidad de Granada.

Murillo-Barroso, M., Montero-Ruiz, I. and Aranda Jiménez, G. (2015) An insight into the organisation of metal production in the Argaric society. *Journal of Archaeological Science: Reports* 2, 141–155.

Oates, D. and Oates, J. (1959) Ain Sinu: A Roman frontier post in Northern Iraq. *Iraq* 21, 207–242.

OXALID (2022) http://oxalid.arch.ox.ac.uk/default.html. Last accessed 12/12/2022.

Paynter, S. (2009) Links between glazes and glass in mid-2nd millennium BC Mesopotamia and Egypt. In A.J. Shortland, I. Freestone and T. Rehren (eds) *From Mine to Microscope: Advances in the Study of Ancient Technology*, 93–108. Oxford, Oxbow Books.

Pernicka, E., Seeliger, T.C., Wagner, G.A., Begemann, F., Schmitt-Strecker, S., Eibner, C., Öztunali, Ö. and Baranyi, I. (1984) Archäometallurgische Untersuchungen in Nordwestanatolien. *Jahrbuch des Römisch-Germanischen Zentralmuseum Mainz* 31, 533–599.

Pernicka, E., Begemann, F., Schmitt-Strecker, S. and Wagner, G.A. (1993) Eniolithic and Early Bronze Age copper artefacts from the Balkans and their relation to Serbian copper ores. *Praehistorisches Zeitschrift* 68, 1–54.

Pernicka, E., Lutz, J. and Stöllner, T. (2016) Bronze Age copper produced at Mitterberg, Austria, and its distribution. *Archaeologia Austriaca* 100, 19–55.

Perreault, C. (2019) *The Quality of the Archaeological Record.* Chicago, University of Chicago Press.

Pollard, A.M., Bray, P., Cuénod, A., Hommel, P., Hsu, Y.-K., Liu, R., Perucchetti, L., Pouncett, J. and Saunders, M. (2018) *Beyond Provenance: The Interpretation of Chemical and Isotopic Data in Archaeological Bronzes*. Leuven, Leuven University Press. https://doi.org/10.2307/j.ctv7xbs5r

Pollard, A. M., Liu, R., Rawson, J. and Tang, X. (2019) From alloy composition to alloying practice: Chinese bronzes. *Archaeometry* 61.1, 70– 82. https://doi.org/10.1111/arcm.12415.

Ponting, M. and Levene, D. (2015) Recycling economies, when efficient, are by their nature invisible: a first century Jewish recycling economy. In M. Geller (ed.) *The Archaeology and Material Culture of the Babylonian Talmud*, 39–65. Leiden, Brill. https://doi.org/10.1163/9789004304895_004

Popper, K. (2002) *The Logic of Scientific Discovery*. 2nd edition. London, Routledge.

Rehren, T. (2001) Aspects of the production of cobalt-blue glass in Egypt. *Archaeometry* 43.4, 13–24.

Rothenberg, B. and Blanco-Freijeiro, A. (1981) *Studies in Ancient Mining and Metallurgy in South-West Spain: Explorations and Excavations in the Province of Huelva*. London, Institute of Archaeo-Metallurgical Studies.

Sainsbury, V.A., Bray, P., Gosden, C. and Pollard, A.M. (2021) Mutable objects, places and chronologies. *Antiquity* 95, 215–227. https://doi.org/10.15184/aqy.2020.240

Salkield, L.U. (1982) The Roman and Pre-Roman slags at Rio Tinto Spain. In T.A. Wertime and S.F. Wertime (eds) *Early Pyrotechnology: The Evolution of the First Fire-Using Industries*, 137–147. Washington D.C., Smithsonian Institution Press.

Salkield, L.U. (1987) *A Technical History of the Rio Tinto Mines: Some Notes on Exploitation from pre-Phoenician Times to the 1950s*. London, Institute of Mining and Metallurgy.

Sherratt, A.G. and Sherratt, S. (1991) From luxuries to commodities: the nature of the Mediterranean Bronze Age trading systems. In N.H. Gale (ed.) *Bronze Age Trade in the Mediterranean: Papers Presented at the Conference Held at Rewley House, Oxford, in December 1989*, 351–386. Jonsered, Paul Åströms Förlag.

Shortland, A.J. (2000) *Vitreous Materials at Amarna: The Production of Glass and Faience in 18th Dynasty Egypt*. Oxford, British Archaeological Reports.

Shortland, A. (2012) *Lapis Lazuli from the Kiln: Glass and Glassmaking in the Late Bronze Age*. Leuven, Leuven University Press.

Shortland, A. (2016) Towards an understanding of the origin of Late Bronze Age glass. In E. Kiriatzi and C. Knappett (eds) *Human Mobility and Technological Transfer in the Prehistoric Mediterranean*, 94–101. Cambridge, Cambridge University Press.

Silvestri, A. (2008) The coloured glass of *Iulia Felix*. *Journal of Archaeological Science* 35.6, 1489–1501.

Silvestri, A., Molin, G. and Salviulo, G. (2008) The colourless glass of *Iulia Felix*. *Journal of Archaeological Science* 35.2, 331–341.

Smirnou, M. and Rehren, T. (2013) Shades of blue – copper-cobalt coloured blue glass from New Kingdom Egypt and the Mycenaean world: a matter of production or colourant source?. *Journal of Archaeological Science* 40.12, 4731–4743.

Stern, E.M. and Schlick-Nolte, B. (1994) *Early Glass of the Ancient World 1600BC-AD50*. Ostfildern, Verlag Gerd Hatje.

Stos-Gale, Z. and Gale, N. (1994) Metals. In A.B. Knapp and J.F. Cherry (eds) *Provenience Studies and Bronze Age Cyprus: Production, Exchange, and Politico-Economic Change*, 92–121. Monographs in World Archaeology 21. Madison, Prehistory Press.

Sun, W., Zhang, L., Guo J., Li, C., Jiang, Y., Zartman, R.E. and Zhang, Z. (2016) Origin of the mysterious Yin-Shang bronzes in China indicated by lead isotopes. *Nature Scientific Reports* 6.1, 23304. https://doi.org/10.1038/srep23304

Sun, W., Zhang, L-P., Guo, J., Li, C-H. and Zhang, Z-F. (2018) Comment on 'did China import metals from Africa in the bronze age? *Archaeometry* 60.5, 1036–1039.

Varberg, J., Gratuze, B. and Kaul, F. (2015) Between Egypt, Mesopotamia and Scandinavia: Late Bronze Age glass beads found in Denmark. *Journal of Archaeological Science* 54, 168–181.

Varberg, J., Gratuze, B., Kaul, F., Hansen, A.H., Rotea, M. and Wittenberger, M. (2016) Mesopotamian glass from Late Bronze Age Egypt, Romania, Germany and Denmark. *Journal of Archaeological Science* 74, 184–194.

Vaxevanopoulos, M., Davis, G., Milot, J., Blichert-Toft, J., Malod-Dognin, C. and Albarède, F. (2022) Narrowing provenance for ancient Greek silver coins using Ag isotopes and Sb contents of potential ores. *Journal of Archaeological Science* 145, 105645.

Wagner, G.A., Pernicka, E., Seeliger, T.C., Öztunali, O., Baranyi, I., Begemann, F. and Schmitt-Strecker, S. (1985) Geologische Untersuchungen zur fruhen Metallurgie in NW-Anatolien. *Bulletin of the Mineral Research and Exploration Institute of Turkey* 101/102, 45–81.

Wagner, G.A., Pernicka, E., Seeliger, T.C., Lorenz, I.B., Begemann, F., Schmitt-Strecker, S., Eibner, C. and Öztunali, Ö. (1986) Geochemische und isotopische Charakteristika fruher Rohstoffquellen fur Kupfer, Blei, Silber und Gold in der Turkei. *Jahrbuch des Römisch-Germanischen Zentralmuseum Mainz* 33, 723–752.

Walton, M.S. and Tite, M.S. (2010) Production technology of Roman lead-glazed pottery and its continuance into Late Antiquity. *Archaeometry* 52.5, 733–759.

Wood, J.R. (2022) Approaches to interrogate the erased histories of recycled archaeological objects. *Archaeometry* 64.S1, 187–205. https://doi.org/10.1111/arcm.12756

Wood, J.R. and Greenacre, M. (2021) Making the most of expert knowledge to analyse archaeological data: a case study on Parthian and Sasanian glazed pottery. *Archaeological and Anthropological Sciences* 13, 110. https://doi.org/10.1007/s12520-021-01341-0

Wood, J.R. and Hsu, Y-T. (2019) An archaeometallurgical explanation for the disappearance of Egyptian and Near Eastern cobalt blue glass at the end of the Late Bronze Age. *Internet Archaeology* 52. https://doi.org/10.11141/ia.52.3

Wood, J.R. and Hsu, Y-T. (2020) Recycling Roman glass to glaze Parthian Pottery. *Iraq* 82, 259–270. https://doi.org/10.1017/irq.2020.9

Wood, J.R. and Liu, Y. (2022) A multivariate approach to investigate metallurgical technology: the case of the Chinese ritual bronzes. *Journal of Archaeological Method and Theory*. https://doi.org/10.1007/s10816-022-09572-8

Wood, J.R. and Montero-Ruiz, I. (2019) Semi-refined silver for the silversmiths of the Iron Age Mediterranean: A mechanism for the elusiveness of Iberian silver. *Trabajos de Prehistoria* 76, 272–285. https://doi.org/10.3989/tp.2019.12237

Wood, J.R., Charlton, M.J., Murillo-Barroso, M. and Martinón-Torres, M. (2017) Iridium to provenance ancient silver. *Journal of Archaeological Science* 81, 1–12. https://doi.org/10.1016/j.jas.2017.03.002

Wood, J.R., Montero-Ruiz, I. and Martinón-Torres, M. (2019) From Iberia to the Southern Levant: The movement of silver across the Mediterranean in the Early Iron Age, *Journal of World Prehistory* 32, 1–31. https://doi.org/10.1007/s10963-018-09128-3

Wreszinski, W. (1931) Das Inhaltsverzeichnis der von Thutmosis III: Erbeuteten Karnak Schätze (The index of Thutmose III. Karnak looted treasures). Leipzig.

Chapter 7

Textile reuse in Roman naval contexts

Margarita Gleba and Maria Stella Busana

Introduction

Textiles were and remain an indispensable part of human existence, being one of the earliest human craft technologies. They have been a fundamental part of subsistence, economy and exchange for millennia. The textile economy in the past had complex production and circulation networks. The acquisition of spinning and weaving skills required a long period of learning and practice. Textile production required substantial resource and time investment. As argued by Susanna Harris (2017), textiles were valued for their material properties, for expensive and/or exclusive raw materials and/or techniques (*e.g.*, silk, shellfish purple, tapestry), for their conspicuous, sensory appeal, for their biography or for their fungibility (substitutability of one object for another). Being highly prized, textiles were used until they lost their original functionality and were then reused, sometimes multiple times.

According to John Peter Wild (2020, 74), the life cycle of a Roman textile comprises six phases: production, primary use, maintenance (including repair), primary reuse, secondary reuse and discard (Fig. 7.1). Examples of primary reuse are clothing items made of different garments, such as trousers or children's tunics made from mantles or adult tunics, or helmet liners made of coarse wool fabrics. Under this category also included are burial shrouds or mummy wrappings.

Under Wild's secondary reuse category are new artefacts created from old rags, the *centoni*, padding for pack animals, padding for defensive armour, *etc.* There are also examples of reuse that fall in-between these two categories. One such item is a tunic from Bernuthsfeld, Germany, dated *c.* AD 660, which is sewn from 45 patches of different weaves and weave qualities. In this case, rags were the raw material for the garment (Grömer 2017, 80–81). Another case is that of the Ottoman period doll dresses from Qasr Ibrim in Egypt (Hitchens 2023).

Fig. 7.1 The life cycle of a Roman textile according to J.P. Wild 2020 (Image by M. Gleba).

The evidence of Roman textile reuse is provided by both written and archaeological sources from across a wide geographical area (Wild 2020; Hitchens 2023). A wonderful example of a certain awareness of a circular economy can be found in Cato's *De Agricultura* (59) where he recommends providing each slave with a new tunic in alternate years, while taking away the old one from which *centones* (patchwork) should be made.[1]

Textile reuse is not a new topic in Roman studies, and much of the discussion in fact revolves around the people called *centonarii.* Cato (Agr. 2, 2) states that the *centonarii* were involved in the textile sector, in particular sewing of discarded textiles or rags. Although, it cannot be excluded that they were also involved in weaving. Renate Lafer (2001) proposed their identification as fire brigades. Jinyu Liu (2009) believes that, even if engaged, together with the blacksmiths, in extinguishing fires, their main activity was to collect rags and textiles used to make *centones*, patchwork cloths made up of many pieces of fabric. These were returned to the textile market and also used in case of fires. She suggests that they may also have been involved in the production of other types of low-priced products, such as military clothing, blankets or animal covers. Their guilds (*collegia*) were concentrated in northern Italy, southern Gaul and the Danubian provinces (Liu 2013). One of the oldest inscriptions mentioning such a guild comes from Padua in Regio X dated before AD 69, probably in the Julio-Claudian period, and mentions an euergetic temple refurbishment financed through a donation by *Marcus Iunius Sabinus, IIII vir aediliciae potestatis, patronus collegii centonariorum* (*CIL* V.2864 = *ILS* 5406).

With respect to the involvement of the *centonarii* in textile production, it is worth mentioning Flavia Solva, an autonomous city in the southeastern territory of Noricum, occupied from the 1st to the 4th centuries AD. The *Rescriptum* of Septimius Severus and Caracalla (14 October 205 AD) addressed to the guild of the *centonarii* of Flavia Solva lists 93 names of these artisans active at that time in the city (Liu 2009, 340–341, n. 39). It is unlikely that the city had that many firefighters. A considerable number of weaving tools at the site furthermore suggests a specialisation in textile production (Gostenčnik 2012, 73–75).

What about the archaeological evidence? In the graph created by Karina Grömer (2017), we can see a wide variety of contexts where textile reuse has been documented across Central Europe in both prehistoric and historic periods (Fig. 7.2). Here we focus on archaeological evidence for Roman textile reuse in naval contexts, which Wild does not mention in his overview of Roman textile reuse, but which is present in Grömer's graph and which – as we aim to demonstrate – is well documented archaeologically.

First, let us review what types of textiles were used in naval contexts that could theoretically be recovered archaeologically. These would have included sails, cordage

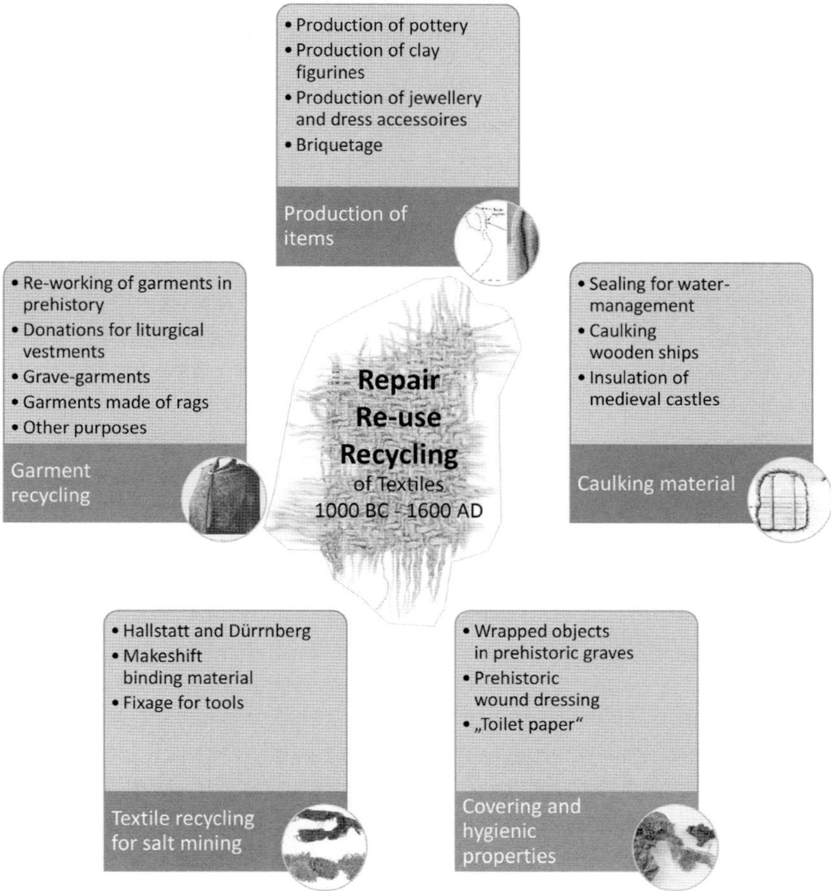

Fig. 7.2 Graph summarising the contexts of textile reuse in Central Europe between 1500 BC and AD 1500 (Image after Grömer 2017, 92, fig. 17, reproduced with permission).

for rigging and other purposes, furnishings such as awnings and screens, clothing for sailors, banners, transport textiles such as sacks and packing, as well as caulking and tools used for tarring or waterproofing (Shaw and Chaplin 2016; Spantidaki 2018). Most of these would have had specific functional requirements and would have been likely produced for these purposes in their primary use. But some of these textiles could have been reused materials, although we cannot be sure, particularly in the context of a shipwreck, where identifying a textile function is not always easy. The exception here is when the textiles are found within the body of the ship – as in the case of caulking, luting and insulation material – and in the case of very specific items such as tarring tools. The study of this type of material is not yet systematic, and we hope that bringing together the emerging data will highlight the importance and the scale of textile reuse in naval and other contexts during the Roman period.

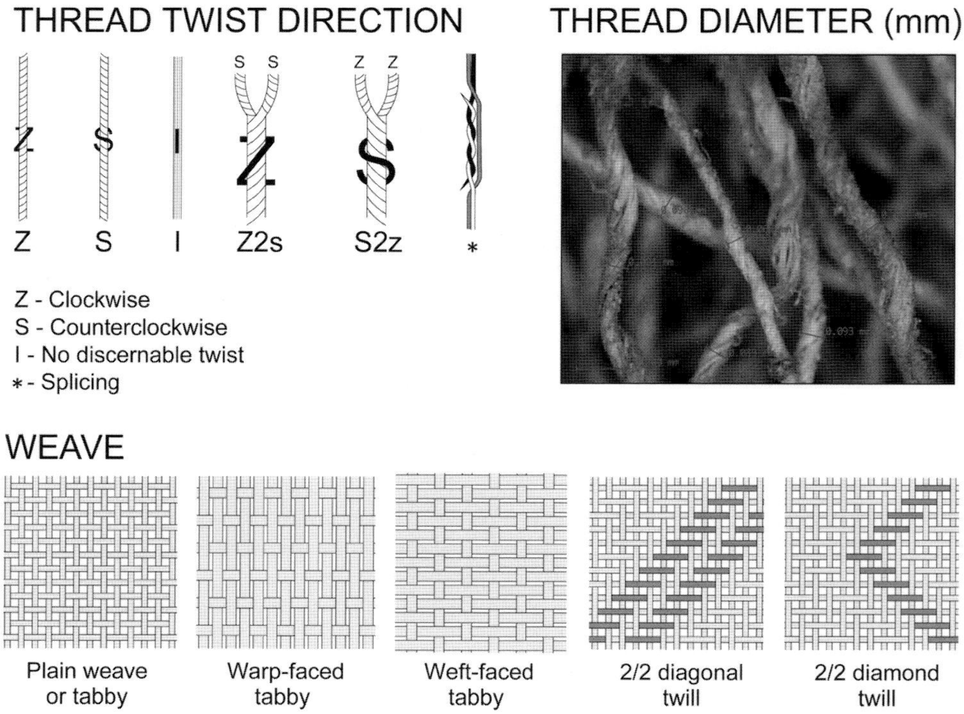

Fig. 7.3 Structural textile elements: thread twist, thread diameter and weave (Image by M. Gleba and V. Herring).

First we will briefly discuss the methodology of textile analysis: the characteristics that may be relevant to textile provenance and function include the origin of the fibre (wool or plant), thread characteristics (structure, diameter, twist direction and angle), thread count per cm in warp and weft, indicative of textile quality, and the type of weave (plain weave/tabby or twill and their variations *e.g.* weft-faced tabby, where the density of the weft is significantly higher than the warp) (Emery 1966; Gleba 2017; Fig. 7.3). In order to identify textile reuse, we will review the structural characteristics of textiles recovered from naval contexts and their specific uses. For example, if the textiles have a variety of weaves, it could be a signal that the textiles were reused and not produced for the specific purpose of insulating ships. We will also present some considerations regarding the quantities of textiles reused in naval contexts based on the surface area of the ships in which they were used, as this type of data is relevant for the reconstruction of the Roman circular economy.

We will concentrate in particular on case studies from Italy and France since they provide the best and most recent evidence. We divided them according to their function: waterproofing; insulating; and textiles used as likely tarring tools for applying pitch to the ship hull (Fig. 7.4). We will focus in particular on the textiles from two Italian contexts, the shipwrecks of Comacchio and Nemi, since they are

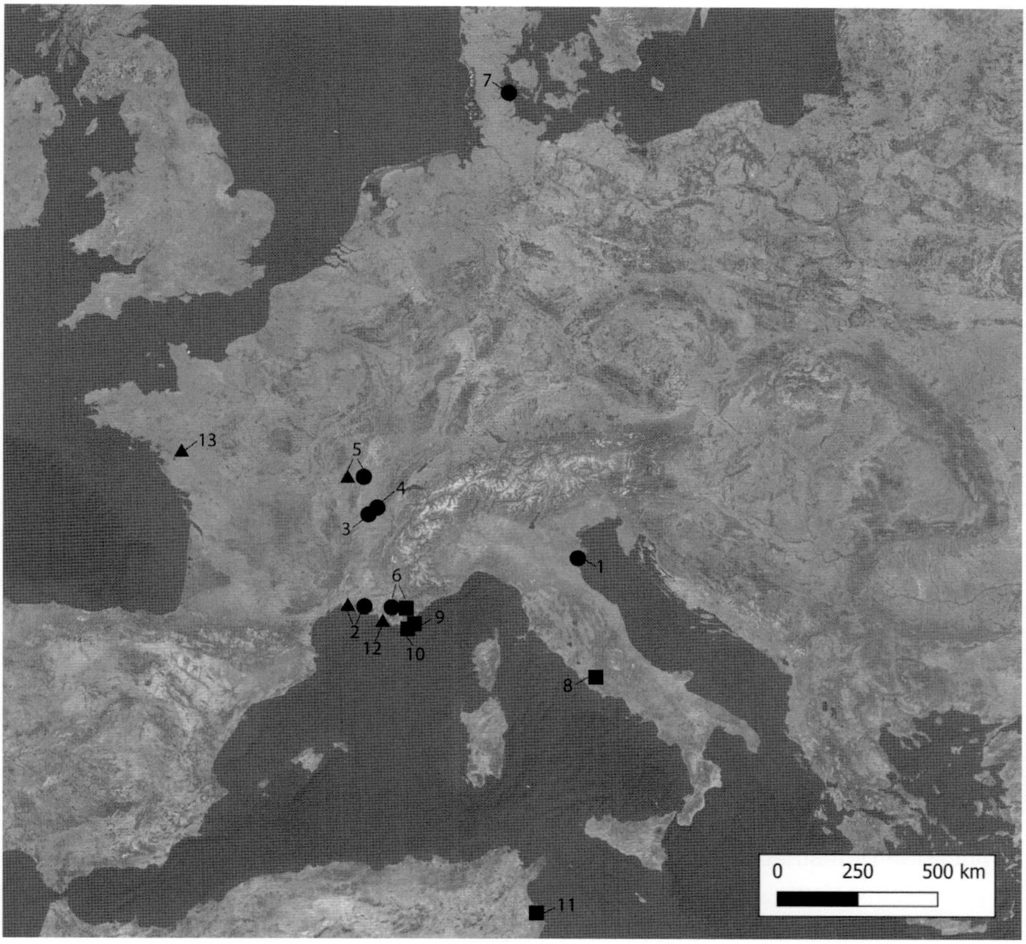

Fig. 7.4 Map showing the location of sites discussed in the text (Image by L. Bernardi). Circle – caulking/ luting; square – insulation; triangle – tarring tools. 1. Comacchio, Italy; 2. Arles-Rhône 3, France; 3. Lyon Saint-Georges 4, France; 4. Tolozan-Lyon, France; 5. Chalon-sur-Saône, France; 6. Grand Congloué-Marseille, France; 7. Nydam Mose, Denmark; 8. Nemi, Italy; 9. Dramont A, France; 10. Madrague de Giens, France; 11. Mahdia, Tunisia; 12. Bourse de Marseille, France; 13. Rezé-Saint-Lupien, France (Image by L. Bernardi).

currently the subject of new research. The discussion on the French material is based on the important and pioneering work carried out by Fabienne Médard.

Waterproofing

We will begin with textiles used in the waterproofing of ships' hulls, that is examples of caulking and luting. Here, we should clarify the difference between the two terms which in the literature are unfortunately used interchangeably, meaning that it is not always easy to understand the exact function. *Luting* is the application of

Fig. 7.5 Comacchio shipwreck: A) view of the ship during excavation (Image after Berti 1990, 25, fig. 3, reproduced with permission); B) view of the hull boards covered with textiles used to seal the caulking between them (Image after Berti 1990, 30, fig. 2, reproduced with permission).

waterproofing material during the construction of the hull and is often associated with clinker construction, more common in North and Central Europe. *Caulking* is the application of insulating material after the hull's construction and is more common in carvel-built ships, common in the Mediterranean. There may also be a difference in the construction of riverine (the majority of our French case studies) and sea-going ships. Furthermore, researchers sometimes disagree as to the technique used. Since it is not the scope of this paper to discuss waterproofing techniques, we will include both terms where the technique is debated.

Our first case study is the Roman ship found and excavated in 1981 in the immediate outskirts of Comacchio, in the Valle Ponti area, north of Ravenna in northeast Italy (Berti 1990; Fig. 7.5A). The vessel was probably wrecked towards the end of the 1st century BC due to a storm. It must have sunk quickly since it retains most of its equipment and commercial cargo, which included 102 lead ingots with seals dated *c.* 12 BC, recently isotopically demonstrated to have originated in Macedonia (Bode *et al.* 2021), six small lead temple models, fineware ceramics of north Italian production, numerous wine and oil amphorae of Adriatic, Aegean and Iberian production, and boxwood logs.

Multiple textile fragments were recovered, of which six were published in 1990 (Castelletti *et al.* 1990), and which are currently being reinvestigated or studied for the first time. At least three of these fragments have cross-shaped imprints of a cord, indicating that they were used to seal the caulking between the hull boards. Thus,

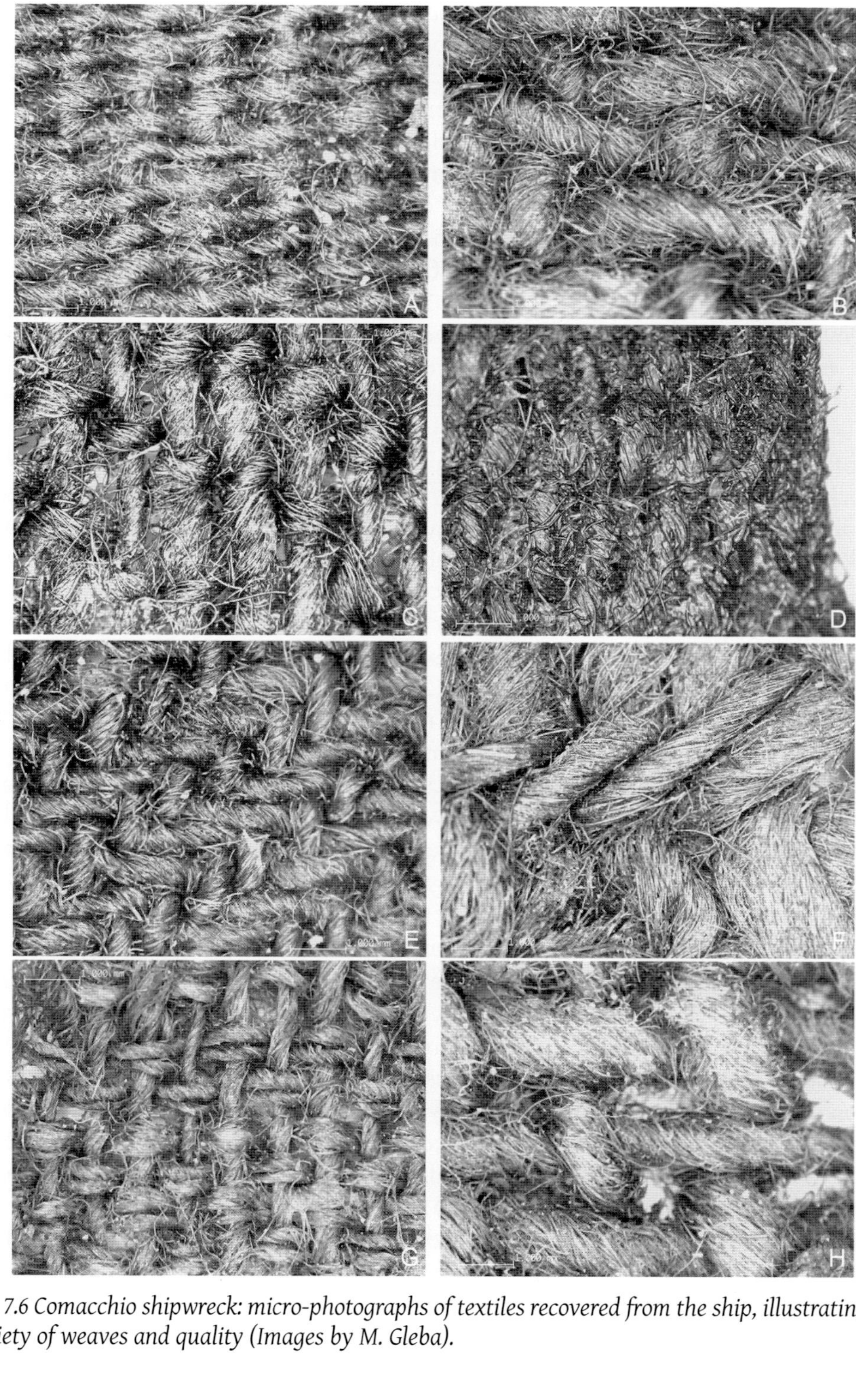

Fig. 7.6 Comacchio shipwreck: micro-photographs of textiles recovered from the ship, illustrating the variety of weaves and quality (Images by M. Gleba).

they can be definitively attributed to a specific use within the ship's construction (Fig. 7.5B). One piece of fabric was interposed between two boards of the ship's hull at the level of the bilge. The function of other fragments is unclear as they were recovered loose among the shipwreck debris.

If we look at the technical characteristics of the Comacchio textiles, we can see a variety of weaves among the contextually attributed fabrics, including different types of tabbies and twills of different qualities (Fig. 7.6). The only commonality between the textiles is that they are all made of wool. Except for one narrow band that could have been a belt or a *strophion*, none of the fragments display any elements that would allow for an identification of their original function.

Moving to France, the excavation of the ancient right bank of the river Saône in Lyon, between 2002 and 2004, revealed 16 wrecks dating from the 1st to the 18th centuries AD. Of these, six are dated to the Gallo-Roman period and all produced textile fragments (Médard 2010). Thanks to the dismantling of the Lyon-Saint-Georges 4 wreck, a flat-bottom boat dated to the 2nd century AD, 26 metres of waterproofing material consisting of tarred fabrics were recovered between the joints of the boards that made up the boat (Meunier and Guyon 2020). Most of the fabric was placed between the bilge and the side. While Fabienne Médard believes the textiles were used for luting, Laure Meunier and Marc Guyon suggest that they were used for caulking, because there is no trace of fabric on the inside. Almost all of the textiles are wool fabrics, although there are also some linen fragments. The weaves include weft-faced tabby, half-basket weave and 2/2 twill of various qualities. Some have been impregnated with ochre. There is a recognisable selection of fabrics based on where they were used: thicker in the corner sector between the bilge and the side and thinner among the sides. Some fabrics are very fine, similar in quality to those used in the Nemi ship, discussed below. While it is difficult to establish what their primary use was, Wild has compared some of these fragments to the finest quality clothing used at Vindolanda, and some fragments were probably sails as they preserve reinforcement strips (Garmi and Meunier 2020, 196).

The Arles-Rhône 3 (AR3) shipwreck was discovered in 2004 on the right bank of the Rhône River in Arles, embedded in the garbage dump of the port area of ancient Arelate (Marlier *et al.* 2017). It is dated between AD 66/70 and 140, with a *terminus ante quem* based on thousands of amphorae and ceramic fragments. The boat was made watertight by means of wool fabrics impregnated with pitch placed between the planks, *i.e.*, caulking. The pollen analyses of the pitch revealed the presence of olive trees, which indicates a local construction for the boat (Marlier *et al.* 2017, 338). Numerous traces of pitch, found both inside and outside the hull, also contributed to the watertightness of the boat. A total of 70 textile fragments belonging to 38 different fabrics were recovered as blocks and studied by Médard (2014). All are wool and include various qualities of weft-faced tabby, half-basket weave and 2/2 twill. Furthermore, there appears to be a 'hierarchy' of qualities used that may reflect a specific use of finer and coarser textiles in particular areas (Médard 2014, 123).

Here we would also like to mention the two boats recovered at Nydam Mose in Denmark in the 19th century. They date to the early 4th century AD and so are roughly contemporaneous with the Roman material from Italy and France. Describing the oak ship B, the excavator Conrad Engelhardt observed in 1865: 'Between the planks where they cover each other, [the boat] is sealed with woollen cloth and a pitch-like and sticky matter.' And concerning the pine vessel C, he wrote: 'Underneath, there are patches of a sealing substance of woollen, woven cloth and a sticky, pitch-like matter' (Möller-Wiering 2011, 93). The textile material was recently analysed by Susan Möller-Wiering (2011) who identified a variety of reused fabrics that were utilised both for the original caulking and for secondary repairs. Textile quality appears to have had little importance to caulking, as is reflected in the high variability of thread counts. The quality of the fabrics is generally similar to contemporary clothing (Möller-Wiering 2011, 94, 96). We can thus see that in the Roman period, textiles were reused as caulking or luting across Europe, even outside the Roman realm, indicating that the technology was quite 'global.'

Insulation

Another practice, more complex and requiring large quantities of fabric, is documented in vessels with a lead lining of the hull, a double planking or both (Médard 2010, 141). Here, between the planking and the lead lining, or between each thickness of planking and lead, fabric impregnated with a pitch-based coating was inserted. This use is attested on the ships of Nemi and the wrecks of the Grand Congloué,[2] Dramont A, Mahdia and Madrague de Giens (Tchernia *et al.* 1978, 85; Médard 2010, 141).[3] One of the hypotheses is that such lining protected the ship from the shipworm and was typical of Hellenic marine ship building (Benoît 1961, 154).

The best example is the exceptional case of Nemi as the textiles from the other wrecks do not survive. Two gigantic ships that belonged to the emperor Caligula (AD 37–41) were recovered in the waters of lake Nemi, southeast of Rome, between 1929 and 1931 (Ucelli 1940) (Fig. 7.7A).[4] The two hulls constitute a *unicum* among ancient ships, not only because of their size but also because they were true floating palaces: the first serving as a residence, the second as a ceremonial ship connected to the nearby Sanctuary of Diana. The ships likely had a ceremonial nature and were intended for the celebration of religious holidays. It has been hypothesised that they served for the *Isidis navigium*, a ritual by which the Romans, when the sailing season resumed on March 5, ingratiated themselves to the goddess Isis, protectress of sailors. Isis was venerated at the Nemi Sanctuary together with Diana. Caligula was killed at the age of 29 and was subjected to *damnatio memoriae*, which erased all traces of his past power; it is likely that the ships suffered the same fate and were stripped of their most precious and easily removable trappings and almost certainly intentionally sunk. Nonetheless, at the time of their excavation, they provided an unprecedented quantity of information regarding Roman ship-building technologies.

Fig. 7.7 Nemi shipwreck: A) aerial view of the second ship during excavation in 1932 (Image: Museo della Scienza e della Tecnologia 'Leonardo da Vinci,' Public domain, via Wikimedia Commons); B) the lining of cloth between wood and lead sheets was visible during excavation (Image: https://simsi.it/la-marea-fontana-natoli/archeologia-subacquea-parte-prima/).

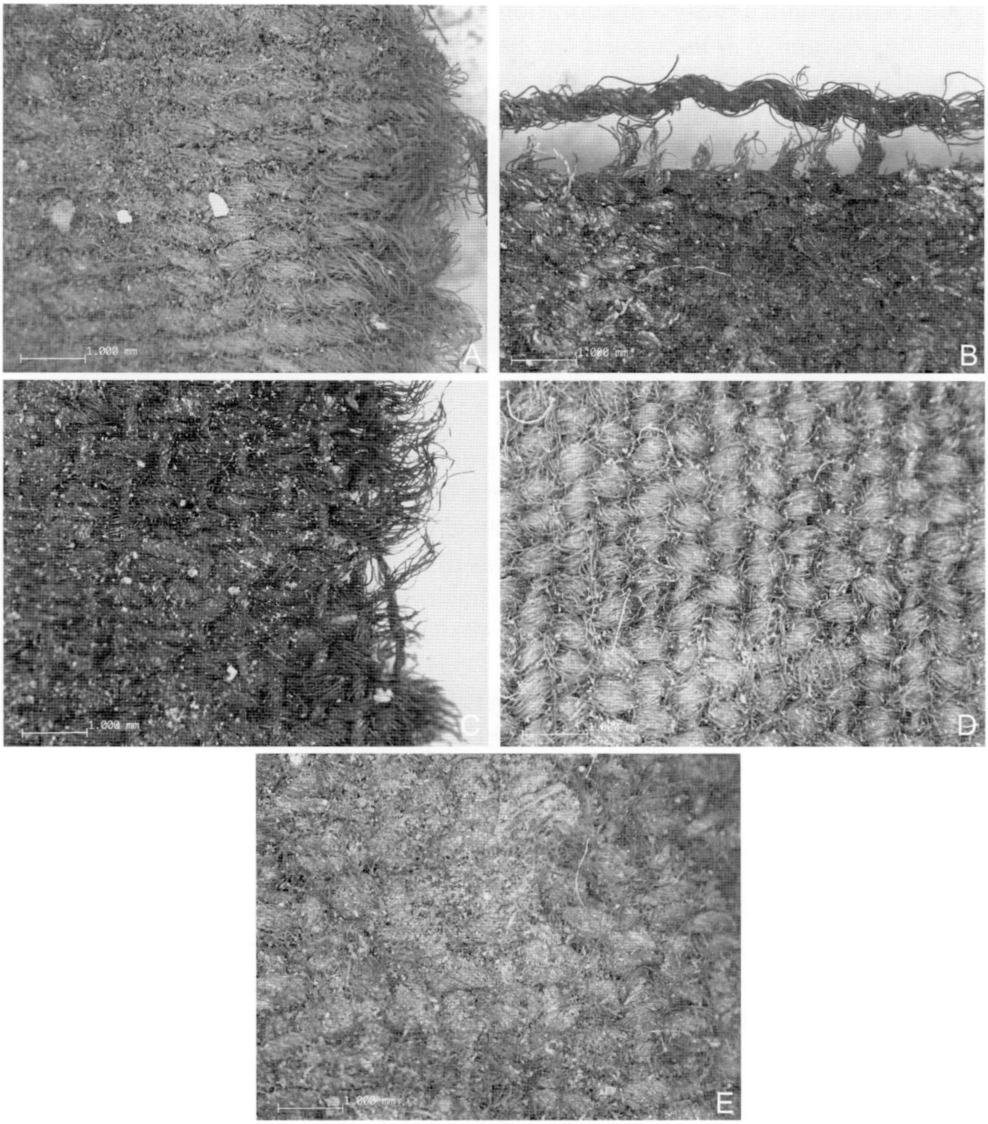

Fig. 7.8 Nemi shipwreck: microphotographs of textiles recovered from the ship, illustrating the variety of weaves and quality (Images by M. Gleba).

Excavation photos clearly show that the external part of the hull was covered with a fabric soaked in impermeable substances, in turn covered with lead sheets held in place by copper nails (Fig. 7.7B). Only four of the textile fragments were studied and published (Levi 1932; Ucelli 1940, 265–268). Unfortunately, the ships were completely destroyed in 1944.

During the reorganisation of the deposit of the Museo delle Navi Romane at Nemi, various unpublished textile artefacts were 'rediscovered', including textile fragments, ropes, cords, and caulking remains recovered in connection with the second ship found in Lake Nemi. Thanks to the collaboration between the Museo delle Navi Romane at Nemi, the University of Padua and the Sapienza University of Rome, a new project was initiated in spring 2022 aimed at understanding their manufacturing techniques, raw materials and function in the naval sector.

The work on the textiles has only just begun, but we can say that the fragments analysed so far are woven in tabby, weft-faced tabby, and half-basket weave, and they differ in quality and appearance (Fig. 7.8). The number of threads per centimetre varies between 10 and 18 in the warp and between 20 and 40 (and up to 60) in the weft. There is also one coarse twill in one of the museum display cases awaiting analysis. All analysed textiles are made from wool.

Tarring tools

Finally, we turn to a collection of finds from French sites that produced textiles impregnated with pitch but without the characteristic crushed appearance of caulking. Instead, they were packaged into balls – called by the French researchers 'boules' or 'galets.' One set of such balls or pebbles was discovered in association with a wreck discovered in 1986 at the Petit-Creusot à Chalon-sur-Saône site, dated to the end of the 2nd or beginning of the 3rd century AD. The analysis revealed that they were impregnated with pine tar of local origin. The presence of tar suggests that these textile objects 'could have been used to coat the hull of the ships with pitch' (Monthiel *et al.* 1998). A total of 42 fragments have been published, ranging in size from 4 to 26 cm. They are primarily of wool, although three fragments are of plant fibre. The most common weave in these fragments is tabby, but twills are also present. Some fragments revealed the presence of dyes.

Textiles collected on the site of the Bourse in Marseille (Bouches-du-Rhône) (1st–3rd centuries AD) presented characteristics similar to those of the Petit-Creusot (Boyer and Vial 1982). They were not directly associated with the wreck and showed no signs of compression typical of caulking (Boyer and Vial 1982, 269).[5] Nine wool and camel hair fragments of different weaves and qualities impregnated with locally produced pitch have been published from this site.

However, it was the discovery of similar objects within the port context of Arles in which the Arles-Rhône 3 shipwreck was found that confirmed the functional interpretation of these textile fragments. Seven more or less complete objects, which resembled small cloth rag dolls or purses enclosing pebbles closed by a ligature, were found in Arles (Médard 2010, 143; 2014 (no. 18); Médard *et al.* 2018; Pl. 12). They had abundant traces of the impregnation of pitch. Most likely they were used in the shipyard to coat the hulls of ships with pitch. The part filled with pebbles was used as a grip, while the loose textile part was used as a sort of brush that was dipped in the pitch. Two of the textile fragments were analysed and include a half-basket weave and a 2/2 twill (Médard *et al.* 2018). The functional interpretation of these objects takes into account their contexts of discovery, which include river ports, shipwrecks (Le Petit Creusot on the Saône, Arles on the Rhône) and larger Roman ports such as Marseille. The association of these textiles with port activity, the presence of pitch on the fabrics and their peculiar shape suggest a link with boat maintenance. The hypothesis that they were used to coat the hulls of boats with pitch is therefore most likely.

The latest similar finds come from the excavation conducted between 2013 and 2016 in the harbour and commercial district of Saint-Lupien in Rezé (Loire-Atlantique), ancient *Ratiatum*. They form part of a *corpus* of 21 textile elements consisting of a total of 26 fragments (Garmi and Meunier 2020). Two of these are 'galets' – compact and homogeneous balls of folded wool textiles impregnated with pitch. Identifiable weaves include balanced tabby, weft-faced tabby and twill. Within this context we should mention three wool twill textiles which the authors identify as fragments of sails based on characteristic stitches or the presence of reinforcement strips (Garmi and Meunier 2020, 194–196).

Discussion

While it is clear how the textiles discussed above were reused, their primary use is often more difficult to ascertain due to their fragmentary state, small size and frequent lack of structural elements that help in functional analysis. The technical characteristics of the fabrics used for caulking, luting, insulation and tarring tools show that the choice of fabric mattered little (see also Médard *et al.* 2018, 505–506). It is this diversity of weaves and qualities that indicates that no standardised production for the purpose was undertaken but rather old rags were reused.

A quick comparison of textile qualities found at all case study sites where the data is available indicates the presence of a wide range of textile types and qualities. The graph on Plate 13 shows thread counts per cm in warp and weft, which is roughly indicative of textile quality. The qualities range between just a few threads per cm in warp and weft to over 40 in warp and up to 100 in weft. In cases where we have statistically larger samples – Arles-Rhône 3 and Lyon-Saint-Georges 4 – the range is wide even within the same weave type. In certain cases, some selection of quality is observed, as at Lyon-Saint-Georges 4. At the same time, there are also some curious differences between different sites, possibly reflecting the context of the textile source. At Comacchio and Marseille Bourse, the textiles are somewhat coarser, while the ships of Nemi produced very high-quality textiles, comparable to those recovered for example at Pompeii (Coletti forthcoming). This may indicate that the reused textiles were sourced locally, at least in some cases. The imperial context of the Nemi ships is thus reflected in the high quality of reused cloth, which possibly originated in the imperial household.

Table 7.1 illustrates the significant variation in weave types. Most of these textile types are well documented in Roman sites across a wide geographical area (*e.g.*, Pompeii (Coletti forthcoming), northern Italy (Busana and Gleba 2018), Vindolanda in the UK (Wild 2012), Egypt (Bender Jørgensen 2018), *etc.*). It seems that the type of

Table 7.1 Summary of data on textiles recovered from various shipwrecks discussed in text (After Médard 2010; 2014; Garmi and Meunier 2020).

Site	*Date*	*No. fragments*	*Weave types*	*Material*	*Reuse*	*Primary use*
Comacchio	1st c. AD	9 analysed	tabby weft-faced tabby half-basket twill	wool	luting/ caulking	garments?
Lyon-Saint-Georges 4	2nd c. AD	95	tabby weft-faced tabby half-basket twill	wool goat 14 plant	luting/ caulking	garments? sails?
Arles-Rhône 3	1st–2nd c. AD	70 (38 fabrics)	tabby weft-faced tabby half-basket twill	wool	luting "galet" (boule)	?
Tolozan sur Lyon	1st c. AD	?	twill	wool	luting/ caulking	?
Nemi	1st c. AD	6 analysed	tabby weft-faced tabby half-basket twill	wool	insulation	garments?
Chalon-sur-Saône Le Petit-Creusot	2nd–3rd c. AD	42	tabby weft-faced tabby warp-faced tabby half-basket twill	39 wool 3 plant	'galet'	?
Rezé-Saint-Lupien	1st–2nd c. AD	26 (21 fabrics)	tabby weft-faced tabby twill	wool	luting/ caulking 'galet'	sails?
Bourse in Marseille	1st–3rd c. AD	9	tabby weft-faced tabby half-basket twill	wool camel	'galet'	garments? sails?

textile did not matter for this function; however, seams, borders and selvedges are few in number in the corpus, indicating that thicker elements of cloth were avoided. Tighter spaces required finer cloth than larger gaps, so a variety of qualities was utilised in naval contexts.

The raw material, on the other hand, appears to have been important in the choice of textiles, indicating that specific material properties were sought for the purpose. The vast majority of the textiles are of animal hair, primarily sheep wool, with some examples of goat and even camel hair, and only a few fragments of plant fibres. The latter are from only two sites, where they are also few and quite small in size. Wool is a material that is very elastic and has excellent absorption properties, important qualities since, in all cases, the cloth was impregnated with resinous substances, such as pitch, tar or bitumen, and then forced into very small spaces (Médard 2014, 123; Garmi and Meunier 2020, 203). It is also possible that plant-fibre textiles were recycled for other functions for which their properties were more suited, such as lamp wicks as documented at Xanthen (Mitschke *et al.* 2012) – a use for which wool would not work – or for mummy wrappings and packing, as in the case of Thebes where fragments of linen sails were reused for the purpose (Wild and Wild 2001, 216).

Although in most cases the original textile function can no longer be determined, in a few cases it is possible to hypothesise about their first use. The qualities, weaves and certain structural or decorative elements of some fragments indicate that they may have been garments, while in two cases it has been suggested that the fragments stem from sails (Garmi and Meunier 2020, 196). Curiously, these fragments are also wool, indicating that at least in the riverine Saône - Rhône boats, wool sails may have been used. This is in contrast to the Mediterranean tradition of using linen.

Finally, the quantities of textiles reused in naval contexts had a direct bearing on their significance for the Roman economy. In order to compare these quantities, the material required has been converted to the amount of fabric needed for an average Roman tunics or toga. For example, we know that the relatively small Lyon-Saint-Georges 4 barge contained at least 26 m^2 of cloth used just for luting/caulking (Garmi and Meunier 2020, 162). This is roughly equivalent to 17 Roman tunics or three Roman togas.[6] Although not all of the textiles were recovered from the Comacchio ship, we can calculate a rough quantity of cloth used to cover the caulking between the planks based on the surviving textiles and documentation: *c.* 20 strips, 22 m long and 7.5 cm wide, that is 33 m^2 of fabric (equivalent to 23.5 Roman tunics or 4 togas). An extremely rough estimate of the surface of the ships of Nemi – the largest Roman ships known – indicates that to cover all of their underwater surfaces with textile insulation would require approximately 1400 m^2 of cloth (equivalent to 1000 Roman tunics or 175 togas) for Ship 1 and 2000 m^2 of cloth (equivalent to 1428 Roman tunics or 250 togas) for Ship 2.

Of course, other ships would have required textiles in quantities somewhere in between these amounts; however, if we take into account the thousands of ships that required fabric either for caulking/luting or for insulation, the need for such textiles would have been substantial. Considering that these textiles had few required functional characteristics and that textile production necessitated substantial resources, time and skills, it would have made sense to reuse old rags for the purpose. At the same time, to ensure the steady and reliable supply of such rags, their collection and distribution must have been somewhat organised. Médard (2014, 124) has suggested that bales of torn rags, sorted by raw material and fabric quality, may have been sent to Roman shipyards. Organisations, such as *collegia centonariorum*, could have had an economic function by managing the collection and redistribution of old textiles on such a large scale. As Liu (2013) suggested, formation of guilds in fact may have been the result of (and a response to) intensified economic activity. Perhaps their formation was also stimulated by large clients, such as the navy, who wanted a single entity with which to do business, or by traders, who wished to obtain public support and authorisation.

This article has demonstrated how a meticulous scientific study of old rags can help glean information about the invisible processes and possible agents of a circular textile economy in the Roman world. Future finds and a re-examination of old ones with new methods and questions in mind will provide statistically more robust data. This is where archaeological science may be of great help, too. Thus, analysis of pollen present in the pitch and the type of pitch can help demonstrate the local origin of shipwrecks (Muller 2005; Marlier *et al.* 2017, 338). Wool quality analysis and isotopic tracing might further help with identifying the provenance of both ships and textiles. Moreover, the analysis of textile characteristics in relation to the possible routes followed by the ship may present a fruitful avenue of research. Comparing textile collections from different contexts may provide even more data: would seagoing ships be more likely to have a wider variety of textiles than river boats or vice versa? And, why? What can this tell us about their supply networks and organisation? As Wild (2020, 83) has pointed out, although Roman society is frequently portrayed as highly consumerist, we can also see a highly practical and 'ingrained "make do and mend" attitude,' and 'The whole topic manifestly deserves closer scrutiny in the future.'

Acknowledgements

We are grateful to Daniela De Angelis (Museo delle Navi Romane di Nemi), Tiziano Trocchi (Museo Archeologico Nazionale di Ferrara), and Laura Ruffoni (Museo Delta Antico in Comacchio) for access to the material. Karina Grömer and Fabienne Médard generously shared their images. MG carried out the analysis of the Nemi textiles and wrote the parts pertaining to the textile finds and their technical characteristics. MSB wrote the parts pertaining to written sources and shipwreck contexts. MG and MSB carried out the analysis of the Comacchio textiles. Both authors co-wrote and equally contributed to the introduction and conclusions.

Notes

1 Cato *De Agricultura* 59: *vestimenta familiae. tunicam P.IIIS, saga alternis annis. quotiens cuique tunicam aut sagum dabis, prius veterem accipito, unde centones fiant.*

2 Benoît 1961, 152. The submerged part of the Grand-Congloué, Marseille of the 2nd–1st century BC was faced with lead. Fine cloth (presumed at the time of discovery to be hempen sailcloth – 'la toile à voile en chanvre englue du une mastique végétale') impregnated with an unidentified product, presumably pine pitch, was interposed between wood and lead sheathing of the keel. In addition, textile caulking of the boards is mentioned. Considering that in the vast majority

of other cases we have wool, the identification of hemp is highly unlikely, but regrettably these textiles do not survive and this information cannot be checked.

3 The merchantman found at Madrague de Giens (1st century BC), one of the largest discovered, sunk for unknown reasons while transporting a large cargo of wine and black glazed pottery from Italy. Here, the lining supposedly consisted of wool textiles.

4 Recently recovered documentation is also available at https://simsi.it/la-marea-fontana-natoli/archeologia-subacquea-parte-prima/ [Accessed 16.02.2023].

5 Boyer and Vial, 1982, 269: 'S'agit-il de chiffons qui ont servi à égaliser la couche de poix protectrice sur des carènes de bateaux et ont été ensuite jetés à l'eau après usage?'.

6 For the calculations we use an average size of *c.* 1.4 m^2 for a tunic and *c.* 8 m^2 for a toga.

Bibliography

Bender Jørgensen, L. (2018) Textiles from Mons Claudianus, 'Abu Sha'ar and other Roman sites in the Eastern Desert. In J.-P. Brun, T. Faucher, B. Redon and S. Sidebotham (eds) *The Eastern Desert of Egypt during the Greco-Roman Period: Archaeological Reports* [online]. Paris, Collège de France. https://doi.org/10.4000/books.cdf.5234.

Benoît, F. (1961) *Fouilles Sous-Marines. L'épave du Grand-Congloué à Marseille.* Paris, Centre national de la recherche scientifique.

Berti, F. (1990) *Fortuna maris: la nave romana di Comacchio - Catalogue de l'exposition présentée au Palazzo Bellini, à Comacchio, du 28 avril au 31 décembre 1990.* Bologna, Nuova Alfa.

Bode, M., Hanel, N. and Rothenhöfer, P. (2021) Roman lead ingots from Macedonia – the Augustan shipwreck of Comacchio (prov. Ferrara, Italy) and the reinterpretation of its lead ingots' provenance deduced from lead isotope analysis. *Archaeological and Anthropological Sciences* 13, 163. https://doi.org/10.1007/s12520-021-01430-0

Boyer, R. and Vial, G. (1982) Tissus découverts das les fuilles du port antique de Marseille. *Gallia* 40, 259–270.

Busana, M.S. and Gleba, M. (2018) Textile production and consumption in Roman *Venetia* (Italy): preliminary results of the study of mineralised fibres and textiles. In M.S. Busana, M. Gleba, F. Meo, A. Tricomi (eds) *Purpureae Vestes* VI, 333–349. Zaragosa, Porticos.

Castelletti, L., Maspero, A., Motella, S. and Rottoli, M. (1990) Analisi tecnologiche dei tessuti. In F. Berti (ed.) *Fortuna maris: la nave romana di Comacchio - Catalogue de l'exposition présentée au Palazzo Bellini, à Comacchio, du 28 avril au 31 décembre 1990*, 157–160. Bologna, Nuova Alfa.

Coletti, F. (forthcoming) *I tessuti di Pompei: materiali, tecniche di lavorazione e contesti.* Roma, Edizioni Quasar.

Emery, I. (1966) *The Primary Structures of Fabrics. An Illustrated Classification.* Washington, D.C., The Textile Museum.

Garmi, D. and Meunier, L. (2020) Les textiles archéologiques romains découverts en contexte nautique et portuaire: les cas de Rezé/Ratiatum (Loire-Atlantique) et de Lyon/Lugdunum (Rhône). *Gallia* 77.1, 187–211. DOI:10.4000/gallia.5513

Gleba, M. (2017) Textiles in Pre-Roman Italy: from qualitative to quantitative approach. *Origini: Rivista di Preistoria e Protostoria delle Civiltà Antiche* 40, 9–28.

Gostenčnik, K. (2012) Austria: Roman period. In M. Gleba and U. Mannering (eds) *Textiles and Textile Production in Europe from Prehistory to AD 400*, 65–88. Oxford, Oxbow Books.

Grömer, K. (2017) Recycling of textiles in historic contexts in Europe. Case studies from 1500 BC till 1500 AD. In I. Miloglav, A. Kudelić and J. Balen (eds) *Recikliraj, ideje iz prošlosti*, 75–98. Zagreb, Arheološki Muzej u Zagrebu/ Filozofski Fakultet Sveučilišta u Zagrebu/ Institut zu Arheologiju.

Harris, S. (2017) From value to desirability: the allure of worldly things. *World Archaeology* 49.5, 681–699. https://doi.org/10.1080/00438243.2017.1413416

Hitchens, S. (2023) *A Diachronic and Holistic Analysis of Fibre Production, Spinning, and Textiles at Qasr Ibrim, Nubia, from the Meroitic Period through to the Ottoman empire (c. 350 BC–AD 1812)*. Unpublished thesis, University of Liverpool.

Lafer, R. (2001) *Omnes collegiati, concurrite! Brandbekämpfung im Imperium Romanum*. Frankfurt am Main, P. Lang.

Levi, C. (1932) Tessuti ricuperati sulle navi di Nemi. *Bollettino della Laniera* 46.3, 156–158.

Liu, J. (2009) *Collegia Centonariorum: The Guilds of Textile Dealers in the Roman West*. Leiden/Boston, Brill.

Liu, J. (2013) Trade, traders and guild (?) in textiles: the case of Southern Gaul and Northern Italy (1st–3rd centuries AD). In M. Gleba and J. Pásztókai-Szeöke (eds) *Making Textiles in Pre-Roman and Roman Times*, 126–141. Oxford, Oxbow Books.

Marlier, S., Poveda, P. and Ranchin, N. (2017) The Arles-Rhône 3 project: from the excavation and raising of a Gallo-Roman barge, in the southern coastal French city of Arles, to its documentation and modelling in 3D (2011–2012). In J. Gawronski, A. van Holk and B. van Tillburg (eds) *Ships and Maritime Landscapes: Proceedings of the Thirteenth International Symposium on Boat and Ship Archaeology, Amsterdam, 2012*, 383–389. Eelde, Barkhuis Publishing.

Médard, F. (2010) Les tissus associés à la construction des chalands gallo-romains du Parc Saint-Georges (Lyon), i[er]-iiie siècles ap. J.-C. In E. Rieth (ed.) *Les épaves de Saint-Georges. Lyon - i[er]-xviiie siècles, Archaeonautica* 16, 136–146. Paris, CNRS Editions.

Médard, F. (2011) Textiles archéologiques et archéologie navale: l'exemple des tissus gallo-romains du Parc Saint-Georges. In G. Boetto, P. Pomey and A. Tchernia (eds) *Batellerie gallo-romaine, pratiques régionales et influences maritimes méditerranéennes*, 105–117. Paris/Aix-en-Provence, Errance/Centre Camille Jullian.

Médard, F. (2014) Les tissus associés à la construction du chaland Arles-Rhône 3: qualité et gestion de la matière première, in Marlier S., Andrieu-Ponel V., Arles-Rhône 3: un chaland gallo-romain du i ersiècle après Jésus-Christ. *Archaeonautica* 18, 116–125.

Médard, F., Djaoui, D. and Kopf, S. (2018) Chiffons Gallo-Romains. Un example inédit de recyclage. In M.S. Busana, M. Gleba, F. Meo and A. Tricomi (eds) *Textiles and Dyes in the Mediterranean Economy and Society*, VI *Purpureae Vestes International Symposium*, 333–349. Zaragoza, Libros Pórtico.

Meunier, L. and Guyon, M. (2020) Waterproofing material from 2nd century AD barge from Lyon: looking for Roman know-how to Caulk with pithed textiles. In M. Bustamante-Álvarez, E.H. Sánchez López and J. Jiménez Ávila (eds) *Redefining ancient textile handcraft. Structures, tools and production processes*, VII *Purpureae Vestes International Symposium (Granada, 2–4 October 2019)*, 161–170. Granada, Universidad de Granada.

Mitschke, S., Leih, S., Becker, P., Meijers, R. and A. Paetz gen. Schieck (2012) Ohne Textilien kein Licht! Untersuchungen zu römischen Lampendochten. In L. Chrzanovski (ed.) *Le luminaire antique: lychnological acts 3: actes du 3e congrès international d'études de l'ILA, Université d'Heidelberg, 21.-26. IX.2009*, 251–262. Montagnac, Éditions Mergoil.

Möller-Wiering, S. (2011) *War and Worship*. Oxford, Oxbow Books.

Monthel, G., Schoefer, M., Valansot, O., de Reyer, D. and Nowick, W. (1998) Les tissus gallo-romains de Chalon-sur-Saône, Étude pluridisciplinaire. *Bulletin du Centre international d'étude des textiles anciens* 75, 21–36.

Muller, S. (2005) Palynological study of Antique shipwrecks from the western Mediterranean Sea, France. *Journal of Archaeological Science* 31, 343–349. doi: 10.1016/j.jas.2003.09.005.

Shaw, M.C. and A.P. Chaplin (2016) Sailing the shining sea: maritime textiles of the Bronze Age Aegean. In M.C. Shaw and A.P. Chaplin (eds) *Woven threads: Patterned Textiles of the Aegean Bronze Age*, 149–181. Oxford, Oxbow Books.

Spantidaki, S. (2018) Investigating maritime textiles in Classical Greece: sails and rigging of the Athenian fleet. In M.S. Busana, M. Gleba, F. Meo and A.R. Tricomi (eds) *Textiles and Dyes in the*

Mediterranean Economy and Society: Proceedings of the VIth International Symposium on Textiles and Dyes in the Ancient Mediterranean World (Padova- Este - Altino Italy 17–20 October 2016), 75–86. Zaragoza, Libros Pórtico.

Tchernia, A., Pomey, P., and Hesnard, A. (1978) *L'épave romaine de la Madrague de Giens (Var): (campagnes 1972-1975): fouilles de l'Institut d'archéologie méditerranéenne.* Paris, Centre national de la recherche scientifique.

Ucelli, G. (1940) *Le navi di Nemi.* Roma, La Libreria dello Stato.

Wild, J.P. (2012) England: Roman period. In M. Gleba and U. Mannering (eds) *Textiles and Textile Production in Europe from Prehistory to AD 400*, 444–450. Oxford, Oxbow Books.

Wild, J.P. (2020) The reuse of textiles in the Roman world. In C.N. Duckworth and A. Wilson (eds) *Recycling and Reuse in the Roman Economy*, 61–88. Oxford, Oxford University Press.

Wild, F.C. and J.P. Wild (2001) Sails from the Roman port at Berenike, Egypt. *The International Journal of Nautical Archaeology* 30.2, 211–220. https://doi.org/10.1006/ijna.2001.0354.

Chapter 8

Functional, spiritual or aesthetic? Investigating reuse in high-status 7th-century necklace pendants from early medieval England

Rowan S. English

Introduction

Reuse and recycling in the past are often discussed as concepts motivated by economic matters; how much reuse and recycling play a part in the Roman economy, for example (*cf.* Duckworth and Wilson 2020). As per Howard Williams and collaborators (Williams and Eckardt 2003; Costello and Williams 2019), I think we can move beyond the economic to think about alternate reasons why reuse or recycling occurred. This is of course connected closely to how the object is valued on a number of different planes (Swift 2020, 426–427). Although reuse and recycling are closely related practices, they can be separated as follows: recycling involves the return of the pre-used object to some kind of raw material status, whereas reuse involves a utilisation of the object for the purpose it was originally made for 'or something else' (Peña 2020, 12). J. Theodore Peña (2020, 12) acknowledges that this definition is not exhaustive, but as I demonstrate here, identifying different 'modalities' for reuse, *i.e.*, the primary motivation for the reuse, will help to clarify thinking on a category that is not as well defined as recycling.

Identifying these object specific modalities of reuse is helpful in moving scholarship beyond the economic, although all three modalities that I will outline in this paper are closely related to value in some sense. Modalities move us away from common economic associations with the term value (Graeber 2001, 1). Of course, recycling can have more than an economic reason; for example, in the modern-day jewellery industry, people inherit jewellery and gemstones and have them remodelled into a piece of jewellery that is more wearable and fashionable. This personally loaded act would, of course, be undetectable in the archaeological record but is still recycling, as the gold, and possibly also the gemstones, are returned to raw material status, while still being imbued with the personal story of the owner. However, I would suggest that the category of reuse provides us with more examples of non-economic or more-than economic modalities, while also providing examples that are more visible in the archaeological record.

In this paper I will use three different case studies to illustrate three different modalities of reuse that could be at work. These modalities are functional reuse, aesthetic reuse and spiritual reuse, all reasons for reuse that are somewhat familiar in the archaeological record of early medieval England (Costello and Williams 2019, 115–116). The following section of the text will further introduce the concept of modalities and how I define each of these modalities. The subsequent sections will then put these modalities into action using examples from the archaeological record of early medieval lowland Britain; more specifically, three different necklace pedants from the graves of 7th-century females. This class of objects has recently been subjected to intensive study, allowing a robust frame of reference (Haworth 2021). Applying the concept of modalities to three objects from the same class allows the diversity of types of reuse within a class to be seen and shows how each object should be considered individually with its modalities.

Defining modalities

In this work, I believe that reuse can be interpreted as having different modalities. I define 'modality' as how the reuse exists and is experienced. This approach moves us beyond the value driven approaches that I outlined above and thus further removes us from simple economic reasons for reuse. There are two methodological qualifications I would like to make about the use of modalities of reuse. First, there can be one or more modalities at work for each different reused object. It is possible that something is used again for practical purposes and for its aesthetic nature. However, as I will illustrate below, I think it is almost always the case that one modality of reuse takes precedence over all the others that may be at work. Second, there is not a finite number of modalities that exist in the world. They are highly contextual, and a new one could be created for each reused object in its context. Below, I will introduce three modalities, before presenting case studies of each in action. I have selected these particular modalities for their broad applicability across many contexts. In essence, each case study demonstrates why this piece of material was selected for reuse over the tens, hundreds or thousands of others that were likely available to communities in the past.

The *functional* modality of reuse is the one that I understand as closely connected to the traditional ideas of economically driven reuse. In this modality, the primary reason for reuse is functional. This does not necessarily include the retention of the original function, but it should operate in an effective manner in its new context. The functional modality does not have to be linked to an intrinsic value, but the object can both have an intrinsic value and be functionally reused. A piece of precious metal, as we will see below, can be reused for function before it is reused for its precious metal value.

An *aesthetic* modality of reuse occurs when a piece of material culture is reused for its aesthetic nature, beyond its function. This can manifest itself in multiple ways, but often the most archaeologically visible are objects where they are clearly designed around a piece of reused material. The reused material in this modality also considers the fact that the aesthetic of some pieces of material culture is unachievable in a different time and space. For example, pieces of material culture that were not able to be made in a particular time or place are prime candidates for aesthetic reuse.

The *spiritual* modality could encompass many things. Crucially, in this context, it involves the perception of an amuletic or apotropaic quality. In short, the object is understood to be imbued with a particular quality which motivates it to be reused. These qualities are not economic, functional or aesthetic, and could stem from multiple different physical properties of the reused piece of material culture. They could be grounded in the material from which the object is made, as materials are often thought to have been imbued with apotropaic meaning (Conneller 2011). They could be grounded in the visual qualities of the material, such as shape or design. This can be distinguished from aesthetic reuse, for example, with a lack of focus on the reused material. It may also be directly connected to the temporality and reused nature of the object; objects with a perceived or unknown past have been argued to be very important in some contexts (Williams and Eckardt 2003). These reasons are not exhaustive, and there could be infinite contextual 'spiritual' aspects about objects that as archaeologists we can no longer access. It could even be argued that all of these could be their own modalities in themselves. However, in this paper, the aim is to show the effective use of the modality as a tool.

Chasing modalities in 7th-century lowland Britain

The 7th century in England was a time of significant changes, with the start of the century looking a lot different to the end. Over the course of the century, the kingdoms became largely Christianised, new forms of settlement like emporia arose and furnished burial declined and eventually ended (Hines and Bayliss 2013, 485; Hamerow 2016; Crabtree 2018, 86–137). All these things indicate that the rhythms of life were in flux; major change was happening. The material used for the case studies below comes from some of the last furnished burials that are seen in lowland Britain (Hines and Bayliss 2013, 485). These necklace pendants come from the graves of biological females, seemingly very high-status individuals within their communities. It has been speculated that these women were royal or some of the first founders or early members of religious houses, or indeed both (Lucy *et al.* 2009; Sherlock 2012, 132; Hamerow 2016). Their societal role has been studied well elsewhere, so I will not reintroduce it here (Hamerow 2016; Haworth 2021). Rather, I note that the reuse discussed in this paper has the contextual backdrop of the elite individual that employed this reuse.

I begin with the reuse modality that may be closest to what would be recognisable as economic reuse, the *functional* modality. The Isle of Ely pendant is brimming with functional reuse. The pendant is made in solid gold, with an openwork gold frame holding a discoid rock crystal bead in the centre. The bead is adorned at the front with four strips of gold with garnet cloisonné decoration in a herringbone pattern, forming a cross motif. At the centre there is a table-cut red-purple stone, whose material is unknown. The pendant has a bail made also of gold that has been applied to the top of the pendant. The cross motif is almost certainly a Christian motif; cross motifs and pendants are not uncommon for this period (Lucy 2016; Pl. 14).

The pendant now resides in the Museum of Archaeology and Anthropology in Cambridge after being purchased for the museum in 1951. It was found some years earlier in a field on the Isle of Ely, Cambridgeshire (Fowler 1948; MAA Cambridge 2023). It has been suggested that this pendant was from the cemetery on the North side of Ely Fields Farm, but to me this seems unlikely, as this cemetery is small and seemingly of 6th-century character and the pendant is more than likely 7th-century (Fowler 1948, 71–72; Lethbridge 1953, 1). It is not a surprise that a high-status, probably Christian, piece of jewellery was present in this locale. Ely was home to a double house monastery that may have grown up into its now-famous cathedral, established by Etheldreda in around AD 673 (Lucy *et al.* 2009, 132). There is also the high-status cemetery of a similar 7th-century date as the Isle of Ely pendant, with the proposed 'princess grave' containing a cruciform pendant (Lucy *et al.* 2009, 88–91). The locale is clearly embedded in both the elite and early Christian spheres.

I would argue that the pendant's four main elements are all likely reused. The bail used to suspend the pendant from the neck, the rock crystal bead that forms the core of the pendant, the garnet cloisonné strip used to form the cruciform design and the central table-cut gemstone. First, it is clear that the bail was not designed for this pendant. The bail is a soldered-on biconical bead rather than a bail that was made purposely for the pendant. The bead is also from the 7th-century; Katie Haworth's (2021) figure 9.148 shows comparable biconical beads. Biconical gold beads correspond to type BE2-a in Hines and Bayliss (2013, 538–540, 573), dating to phase AS-FE, corresponding to the middle decades of the 7th-century. The bail broadly falls into Haworth's type three for bails that were probably constructed in the same way as biconical beads (Haworth 2021, 71). I would argue this group could be divided into two subgroups. The first subgroup is those made like biconical beads but specifically for the pendants, where the biconical bead-like decoration ends on the front and has a pulled finished to the back like Haworth's type one and two suspension loops (Haworth 2021, 70). The second group is where the Isle of Ely pendant would fit, where the suspension bails are actually reused biconical beads. We can see this due to the soldering of the bead to the pendant and the crushing of some of the garnet cloisonné cells by the bead; this was clearly not made for this pendant. We can see the reuse in the functional modality as the size of the bead is rather large in proportion to the rest of the pedant, almost dwarfing it; it is not the most aesthetically pleasing and draws the eye away from the gemstones of the pendant. Suspension loops in 7th-century pendants are usually a lot smaller and in proportion, as shown in the two other pendants considered in this paper (Pls. 15 and 16). The reuse was then functional, to hang then pendant with an object that was to hand, not to be aesthetically pleasing. The bead was of course made to be aesthetically pleasing, but its poor aesthetic employment illustrates the functional reuse.

It could also be more tentatively argued that the four garnet cloisonné strips that form the cross arms are reused. The mounting of these garnet cloisonné onto gold strips seems unusual. This mounting could be to shore up the reused material to make it more sound for mounting. However, these strips could also be to add stability to the mounting, irrespective of whether the cloisonné was reused. The quality of the cloisonné is also very good in this pendant. This good quality could also suggest it has been reused as it does not seem to be in agreement with the clumsy soldering and use of a less-than-perfect bail for the object.

The central stone was also likely a reused element. The stone is a table-cut gem of red-purple colour. While the table-cut gem is not unknown in other 7th-century pendants, cabochons are a much more common type of gemstone. These table-cut stones have often been proposed to be seal blanks of Sassanian origin (Haworth 2021, 42). However, the one from Ely is a little different than other 'seal blank' pendants from 7th-century necklaces, as the stone has more of a purple colour rather than a classic red garnet. The circular shape is also uncommon. The closest comparable item I could find for the flat-topped circular stone is the stone set in the pectoral cross of the Staffordshire hoard (Fern *et al.* 2019, 110). The functional reuse could technically be the reuse of the 'seal blank' that forms the centre of the pendant, if this arrived in England as an unmounted seal blank. However, because of the difference between this stone and the other 'seal blanks' in the 7th-century corpus, the case for it being an unmounted blank is unconvincing. There is evidence for jewellery traveling long distances in this period (Haworth 2021, 42), so I would postulate that this flat-topped stone arrived in England in a different piece of jewellery and was then functionally reused in the Isle of Ely pendant. I am not convinced that this stone is a seal blank. It may be the case that the stone was recut from cabochon to table cut.

The final reused element of this pedant is the rock crystal bead that forms the main body of the pendant. Rock crystal beads are common in early medieval graves, often dating to the second half of the 6th century, but they are also known to appear in cremation graves of the 5th century (Guido 1999, 2; Hills and Lucy 2013). The bead here is of discoid shape, so likely to be 6th-century. This bead was then likely an object that survived in circulation until the 7th century, when the functional elements were put together by a craftsperson to form a piece of jewellery fitting for the time. The totality of the reused elements in this pendant actively reenforces the fact that each of the other elements are of a reused nature. The four functionally reused elements come together to form a (most probably) piece of early Christian jewellery, very fitting for its time.

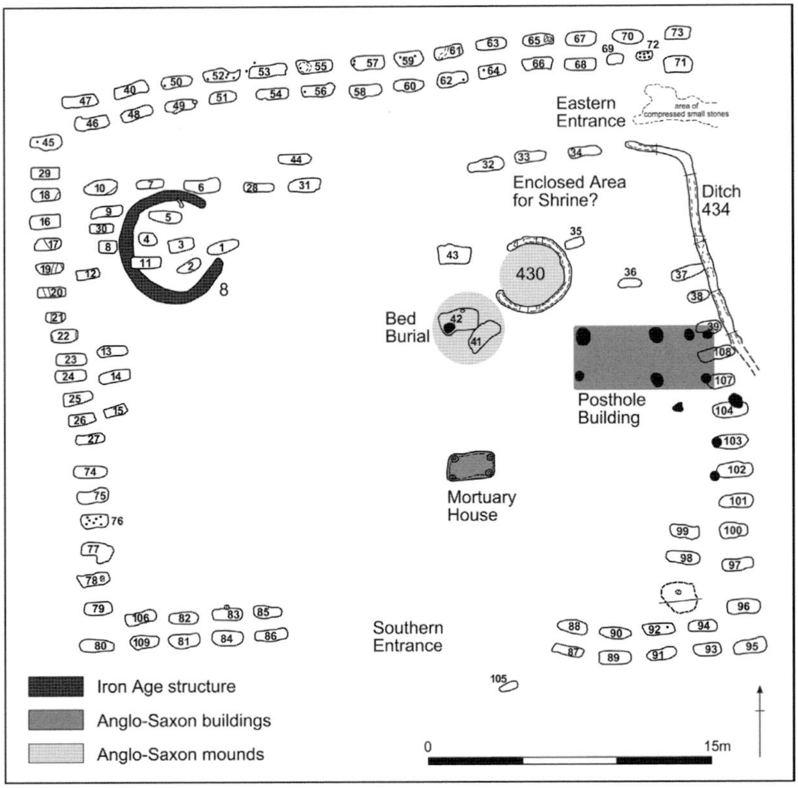

Fig. 8.1 A plan of the site at Street House, Loftus, Northeast Yorkshire. Graves 42 and 43 at the centre of the site are those discussed in this article. Image from A Royal Anglo-Saxon Cemetery at Street House, Loftus, North-East Yorkshire *by Stephen J Sherlock. (Sherlock 2012, fig. 6.3).*

The aesthetic modality moves beyond the functional, with the primary reason for reuse down to the aesthetics. Of course, some things used in the functional modality will be aesthetically pleasing, but this is not always the primary reason for reuse. This difference is made clear below when we see the pendant designed around the aesthetic nature of a central element. The 'princess grave' (grave 42) from Street House is one of the most recently published graves of the 7th-century elite female type (Sherlock 2012). It comes from the centre of a highly organised cemetery where rows of graves form an enclosed space, and where grave 42 sits at the centre, almost paired with grave 43. The site even has some evidence of cobbled surfaces at the entrances to the site (Sherlock 2012, 127). Grave 43 also contains a necklace with a reused element, which will be discussed as our final case study. Figure 8.1 shows the site plan of Street House, with the position of the graves indicated.

Returning to grave 42, the pendant is one of garnet cloisonné in a shield shape with a central shell-shaped, carved red garnet. The pendant again likely dates from the mid-7th century (Sherlock 2012, 48–50; 2016). The piece of reused material is the central shell-shaped garnet, which is most likely Roman. If it was carved in the early medieval period, it would be 'a unique piece of workmanship for this [the 7th century] time' (Sherlock 2016, 250). The craftspeople in this period likely then did not have the skill to create such a piece of jewellery. This unique piece of Roman material culture was then redeployed in the aesthetic modality and has a pendant designed around it (Pl. 15).

There are several clues in the design and manufacture of the pendant to suggest that it has been designed for the central stone, most notably, the lack of symmetry in the garnet cloisonné. This is particularly noticeable on the shoulders of the shield, where the cloisonné panels end in different places because of the available garnets. The garnets in this pendant were used to accommodate the stone and its aesthetics. This suggests that the garnets forming the cloisonné were also reused, and this is reinforced by the variation in colour and size. These inconsistencies in design and craftspersonship are not seen in some of the highest-status cloisonné metalwork of the Staffordshire hoard or from Sutton Hoo. The garnets are also at differing depths in their settings creating an uneven surface, suggesting that the garnets came from different pieces originally (Sherlock 2012, 48–50; 2016). Stephen Sherlock suggests that the shell iconography is a Christian mechanism due to its close association with the bed burial and its Christian connotations (Sherlock 2016, 250–251; *cf.* Brownlee 2022). I am not convinced that the shell-shaped garnet is being employed as a Christian motif in this context. The 7th century was a time before the key Christian connotations of the scallop shell had been formed (Sherlock 2016). I think it is more likely that the shell is an obvious example of the aesthetic reuse modality at work. The likely reason the shell-shaped garnet was reused because this aesthetic was unobtainable in the 7th century.

The final modality of reuse moves us even further away from the simple economic reason for reuse into the category I have defined as the 'spiritual' modality. In this category, the object is reused not because of its practical or aesthetic use but for something *more.* This could be something related to the material nature of the object, the circumstances of the procurement or obtainment of the object, or indeed a visual characteristic of the material. There are many examples of this kind of reuse, especially in early medieval England. Audrey Meaney's *Anglo-Saxon Amulets and Curing Stones* (1981) cites a wide range of materials that were amuletic and apotropaic, with many of these actually being reused material, for example, the Roman jet bracelet bead used as a pendant at Castle Bytham, Lincolnshire (Meaney 1981, 74).

Of course, this is the hardest modality for us, as viewers of the past, to access. I think there is an argument to be made that the reuse of the Iron Age glass bead in grave 43 at Street House could be placed in the spiritual modality. Grave 43 is the second 'focal' grave of the site, paired with the 'princess grave' (42). The necklace pendant here is again what gives us the reused element. The centre plaque of the pendant is part of an Iron Age bead of the Colchester 6a type. The plaque has been cut from the face of the bead making the most of its striking design and the unusual undulating surface (Sherlock 2016). This may suggest that the object is one for multiple senses:

a feast for the eyes of the viewers, but also for the wearer to run their fingers over the undulating surfaces and appreciate the object. The bead has been mounted in a very typical gold mount of the 7th century (Pl. 16). This on the surface is a very straight forward practice, someone is in possession of an aesthetic object that is then made into a pendant. However, I think there are several aspects that could be postulated to move us on from the aesthetic modality to the 'spiritual' modality.

First, I think the fact that the reuse of the specific Colchester type 6a beads happens on a further two occasions reinforces the suggestion that these are likely reused for more than their aesthetics. Along with the pendant from Street House, there are pendants from Cow Lowe, Derbyshire and from Sheffield's Hill, Lincolnshire that are made of Colchester 6a beads (Meaney 1964; Sherlock 2016). These pendants all share some design sensibilities, with the parts of the bead used being carefully selected. They are all set in a manner to show three of the characteristic white spirals on the bead, arranged in a triangular pattern (Haworth 2021, 213). They are also all from northern cemeteries (Haworth 2021, 53). Does this mean that all of the pendants were made from the same bead type that was imbued with something that made its motivation for reuse spiritual, or are the pendants referencing each other (Haworth 2021, 213)? I think this referencing of each other can be further supported by another pendant that may be a copy of the three Colchester 6a examples. The blue and white glass cabochon in silver mount from North Cave, Everthorpe, East Yorkshire has some visual affinities with the Colchester examples (Lucy 1999, 37; Hull City Council 2022). The cabochon here has no evidence of being an adapted bead, making it possible that it was made to directly emulate the Colchester group. This proposal is supported by the fact that it falls strongly within the northern distribution area of these pendants (Haworth 2021, 53). I think the setting of the pendant from North Cave in a lesser metal, silver rather than gold, supports the idea that this pendant is a lesser emulation of the tradition.

Second, the spiral motif that is seen on the glass bead is one that would have been very familiar from nature to those in the early medieval world, as well as one that was born out in negotiations with nature in activities such as the weaving of a basket (Ingold 2000, 431). If this bead was found in the ground from a not-obviously anthropogenic context, could it have been seen as something natural? There was the skill in early medieval England to make polychrome glass beads that look very similar to the Iron Age Colchester 6a beads (see Guido 1999, pl. 5, beads 6x and 6xv), but it is not these beads that are reused. It is the Iron Age type that is seemingly specifically selected. If the object was plucked from an Iron Age grave, this may have meant that it was imbued with *something*, the knowing of its past promoting its reuse. Or is it possible that the spiral motif and its appearance in nature made the finder think that this bead plucked from the ground was something from nature and thus imbued with *something* from its natural origins?

Haworth (2021, 53) characterises this small tradition as a 'complex trajectory from the point of recovery to the point of recontextualisation.' I agree with this sentiment put forward by Haworth. I think it is plausible that there is referencing between these three or four pendants. Perhaps we can even take Alexandra Pesch's (2007, 44–49) idea of *Formularfamilien* from her study of Bracteate where there are 'groups that are so alike they could not have been designed and crafted independently' (Behr *et al.* 2014, 57). I think this is a perfectly plausible concept to apply to this group of pendants.

Do these pendants represent three different Colchester 6a beads used as 'raw' material for these pendants? Or was a single bead type used to make these pendants by the same craftsperson? If so, it is possible that these bead(s) pendants were seen on one of these elite individuals and emulated in a cut-price version with a manufactured glass cabochon set in silver rather than gold with an original bead? They were worn and later taken out of circulation by being placed in graves with individuals around AD 650. I argue the spiritual modality can clearly be seen here. There are too many sings and coincidences within this group to limit them to the aesthetic modality. The beads' reuse could be seen to be referencing a shared elite identity through common practice, or a connection or ability to harness nature for elite power.

Conclusion

Reuse happens for different reasons, many beyond economic sensibilities. In this work I have introduced some alternate modalities for reuse, with examples from the elite female graves of 7th-century lowland Britain. The functional modality of reuse is the one aligned most closely with economic sensibilities but works to show that things can be reused for a practical reason beyond just their value. This practical functionality was illustrated in the case study of the Isle of Ely pendant where we see four elements of different materials come together for practical reason to form a piece of jewellery of its time.

The aesthetic modality of reuse works to foreground the reused object because of its aesthetically pleasing nature. The object is planned around the reused object, often because it represents something that could not be made by the community who is reusing it. This creates an otherwise unachievable aesthetic beauty that was likely very attractive. The aesthetic modality can be seen in the Street House scallop pendant from grave 42. The pendant was designed around the shell-shaped central garnet giving it its unique shield shape. The other elements, while probably being reused, only seek to reinforce the aesthetic modality, as they do not draw attention away from it but work to enhance it.

The final 'spiritual' reuse modality moves us further from the economic and introduces us to some amuletic or apotropaic motivations for reuse. The reuse could stem from a visual quality of the reused material, something innate that the material has or indeed the circumstances in which it was passed to the current owner. This 'spiritual' modality was well illustrated well by the Street House grave 43 Iron Age glass bead pendant and its small family of comparable pendants scattered across northern England in the 7th century. These beads were clearly carefully selected and reused for a reason that for us may never be truly known. However, there is some evidence for the reuse being associated either with nature, elite power, or the beads' history (or lack thereof).

Reuse and curation were clearly practices that would have been well known to the people living in 7th-century lowland Britain (Lucy 1999; Costello and Williams 2019; Werthmann 2019; Swift, 2020; Haworth 2021, 51–54) but are often understood as simply amuletic in the grave or for economic and value-based reasons. The introduction of the modalities above, as I have demonstrated, brings additional clarity to the reasons driving reuse. This newfound clarity shows us that each object has its own unique set of modalities that motivate its reuse, but with one modality seemingly always taking precedence.

Acknowledgements

This work was supported by the Danish National Research Foundation under the grant DNRF119 - Centre of Excellence for Urban Network Evolutions (UrbNet). I would like to thank Katie Haworth for providing unpublished literature and Stephen J. Sherlock for providing the images from Street House.

Bibliography

Behr, C., Hines, J. and Pessell, T. (2014) The Bracteate Hoard from Binham. *Medieval Archaeology* 58, 44–77.

Brownlee, E. (2022) Bed burials in early medieval Europe. *Medieval Archaeology* 66.1, 1–29.

Conneller, C. (2011) *An Archaeology of Material. Substantial Transformation in Early Prehistoric Europe.* Abingdon, Routledge.

Costello, B. and Williams, H. (2019) Rethinking heirlooms in early medieval graves. In M.G. Knight, D. Boughton and R.E. Wilkinson (eds) *Objects of the Past in the Past: Investigating the Significance of Earlier Artefacts in Later Contexts*, 115–129. Oxford: Archaeopress.

Crabtree, P.J. (2018) *Early Medieval Britain: The Rebirth of Towns in the Post-Roman West.* Cambridge, Cambridge University Press.

Duckworth, C.N. and Wilson, A. (2020) Introduction: recycling and reuse in the Roman economy. In C.N. Duckworth and A. Wilson (eds) *Recycling and Reuse in the Roman Economy*, 1–8. Oxford, Oxford University Press.

Fern, C., Dickinson, T. and Webster, L. (2019) *The Staffordshire Hoard. An Anglo-Saxon Treasure.* London, Society of Antiquaries of London.

Fowler, G. (1948) Cratendune: A problem of the Dark Ages. *Proceedings of the Cambridge Antiquarian Society* 41, 70–73.

Graeber, D. (2001) *Toward an Anthropological Theory of Value. The False Coin of Our Own Dreams.* New York, Palgrave.

Guido, M. (1999) *The Glass Beads of Anglo-Saxon England.* M. Welch (ed.). London, Society of Antiquaries of London.

Hamerow, H. (2016) Furnished female burial in seventh-century England: gender and sacral authority in the Conversion Period. *Early Medieval Europe* 24.4, 423–447.

Haworth, K. (2021) *'Most Precious Ornaments' Necklaces in Seventh-Century England.* Unpublished thesis, Durham University.

Hills, C. and Lucy, S. (2013) *Spong Hill. Part XI: Chronology and Synthesis.* Cambridge, McDonald Institute of Archaeology.

Hines, J. and Bayliss, A. (2013) *Anglo-Saxon Graves and Grave Goods of the 6th and 7th Centuries AD: A Chronological Framework.* Abingdon, Routledge.

Hull City Council (2022) *KINCM:1958.83.1 - Hull Museums Collections.* Available at: http://museumcollections.hullcc.gov.uk/collections/search-results/display.php?irn=22035&keywordsorig=blue+glass+pendant&titleorig=&personorig=&placeorig=&dateorig=&materialorig=&accessionnumberorig=&collectionorig=&museumorig=&keywords=anglo+saxon+pendant+&SearchSubmit_x=0&SearchSubmit_y=0&newsearch=new&title=&person=&place=&date=&material=&accessionnumber=&collectionall=all&museumall=all&location=any&Sender=List&Page= (Accessed: 20th February 2023).

Ingold, T. (2000) On weaving a basket. In T. Ingold (ed.) *The Perception of the Environment: Essays on Livelihood, Dwelling and Skill*, 426–438. London, Routledge.

Lethbridge, C. (1953) Jewelled Saxon pendant from the Isle of Ely. *Proceedings of the Cambridge Antiquarian Society* 96, 1–3.

Lucy, S. (1999) Changing burial rites in Northumbria AD 500–750. In J. Hawkes and S. Mills (eds) *Northumbria's Golden Age*, 12–43. Stroud, Sutton.

Lucy, S. (2016) The Trumpington Cross in context. *Anglo-Saxon England* 45, 7–37.

Lucy, S., Newman, R., Dodwell, N., Hills, C., Dekker, M., O'Connell, T., Riddler, I. and Walton Rogers, P. (2009) The burial of a Princess? The later seventh-century cemetery At Westfield Farm, Ely. *The Antiquaries Journal* 89, 81–141.

MAA Cambridge (2023) *Object Record - 1951.565, MAA Cambridge Collections search.* Available at: https://collections.maa.cam.ac.uk/objects/424885 (Accessed: 18 February 2023).

Meaney, A. (1964) *A Gazetteer of Early Anglo-Saxon Burial Sites.* London, George Allen and Unwin Ltd.

Meaney, A. (1981) *Anglo-Saxon Amulets and Curing Stones.* Oxford, British Archaeological Reports.

Peña, J.T. (2020) Recycling in the Roman world: concepts, questions, materials, and organization. In C.N. Duckworth and A. Wilson (eds) *Recycling and Reuse in the Roman Economy*, 9–58. Oxford, Oxford University Press.

Pesch, A. (2007) *Die Goldbrakteaten der Völkerwanderungszeit - Thema and Variation.* Berlin, De Gruyter.

Sherlock, S.J. (2012) *A Royal Anglo-Saxon Cemetery at Street House, Loftus, North-East Yorkshire.* Hartlepool, Tees Archaeology.

Sherlock, S.J. (2016) Use of 'antiques' in conversion period cemeteries. *Medieval Archaeology* 6.2, 242–265.

Swift, E. (2020) Reuse of Roman artefacts in late antiquity and the early medieval west: a case study from Britain of bracelets and belt fittings. In C.N. Duckworth and A. Wilson (eds) *Recycling and Reuse in the Roman Economy*, 425–446. Oxford, Oxford University Press.

Werthmann, I. (2019) *The Past in the Past: the Reuse of Roman Objects in early Anglo-Saxon Society c. AD 400 - c. 700.* Unpublished. thesis, Durham University.

Williams, H. and Eckardt, H. (2003) Objects without a past? The use of Roman objects in early Anglo-Saxon graves. In H. Williams (ed.) *Archaeologies of Remembrance. Death and Memory in Past Societies*, 141–170. London, Plenum Press.

Chapter 9

Stars aligned: tracking the use and reuse of Viking Age metal-casting models for star-shaped brooches through 3D visualisation

Derek Parrott

Introduction

With an increased focus on early medieval circularity over the past several decades, it seems only appropriate that the Viking Age in Scandinavia (AD 750–1100) be given its due academic course in this context as well. Though circular processes of reuse and recycling have been discussed in archaeological terms for decades (for operating definitions, see Schiffer 1987, 27–46), scholarly discussions of these concepts in the Viking Age are rather fragmentary and largely focus on the reuse and recycling of monuments and single objects (see, for example, Pedersen 2006; Andrén 2013; Klevnäs 2016; Lund and Arwill-Nordbladh 2016; Parrott 2020; Lund 2022). What is needed for a more holistic view of this period are investigations of those processes termed 'invisible circularity', meaning those that cannot simply be observed autopically or by otherwise more traditional methods of research. Typically, these processes are those that involve some measure of destruction, repair, or regeneration, such as the production of glass or metal, in which more laboratory and/or bulk finds analysis is needed (Bavuso *et al.* 2023, 12–13). Such analyses have been conducted in various ways in investigations of past circularity, whether that be the study of Roman glass recycling through chemical analysis (Peña 2020, 25, for an overview) or holistic bulk waste material quantification, such as the 2021 study of recycling during systemic collapse at the Mayan site of Ucanal, Guatemala, conducted by Christina Halperin (2021). In the field of Viking Age studies, the theme of invisible circularity has been less explored, though recent investigations of workshop waste from the production of both nonferrous metal goods and glass beads from Ribe (using micro XRF and isotopic analysis, respectively), for instance, has demonstrated the ability of more technical methods to detect raw material recycling and large networks of circularity (Orfanou *et al.* 2021; Barfod *et al.* 2022).

This paper will address another aspect of the nonferrous metalworking industry from this perspective: the use and reuse of casting models to produce nonferrous (namely copper-alloy, but also silver, gold and lead) goods in the so-called emporia of Viking Age Scandinavia, such as Birka in Sweden, Kaupang in Norway, or Ribe in Denmark. While it has been well-established that a great quantity of models for various forms of jewellery, amulets and fittings were reused to serially produce these objects in the emporia, they only rarely appear in the archaeological record (Drescher 1983; Brinch Madsen 1984; Fuglesang 1987; Feveile 2002; Maixner 2005; Pedersen 2016, 38; 2017). As their design would have been the first practical step in the production process of a cast object, being able to trace these models would be fundamental in establishing new understandings of object chronologies and development, crafting and distribution networks, and the circular processes that drove them. There are thus several questions that need to be answered about the nature of their reuse:

- How often were models reused before being discarded?
- How dramatically did objects change from model to model?
- Why might a model be discarded?

Some of these questions have been posed before, particularly with regard to how often models were reused (Pedersen 2016, 70). Based on experimental archaeology, Unn Pedersen (2016, 70) estimated that a lead model, for instance, could be reused hundreds of times. However, more materially grounded studies are needed. A key source to further rectify these issues are the comparatively large amounts of clay mould fragments that have been unearthed by the thousands in the emporia workshop layers (Brinch Madsen 1984; Feveile 2002; Ambrosiani 2013; Pedersen 2016; Sindbæk 2022a). These moulds, used to cast the final objects, contain impressions produced using the now absent models and thus represent a contextual record of these often-ephemeral objects (Brinch Madsen 1984, 91). Due to the nature of the production process (detailed further below) the moulds are heavily fragmented, scorched, and contain only negative impressions, making them a difficult material to discern autoptically (Pl. 17). However, new advances in archaeoscience means that methods are now available that can, quite literally, shed light on these inscrutable objects, in the form of 3D scanners as a tool for image reconstruction and comparison (see for instance, Croix *et al.* 2019; Croix *et al.* 2020; Deckers *et al.* 2021). This paper will apply this method to a small sample of newly excavated mould fragments from the recent 2017–2018 Northern Emporium excavations jointly undertaken by Aarhus University and the Museum of Southwest Jutland at the early Viking Age emporium of Ribe in southwestern Denmark (see Sindbæk 2022a). The goal here is twofold: to trace the models of an individual object type and to demonstrate the viability of this new visualisation method more broadly for illuminating a largely invisible type of material.

Casting Ribe in the Viking Age

Based on both written sources and more than 50 years of archaeological excavations, Ribe has been called Denmark's oldest town (Bencard and Jørgensen 1990; Feveile and Jensen 2000). Located on the southwest coast of the Jutland peninsula, Ribe has its roots in the early 8th century, near the dawn of the Viking Age (Bencard and Jørgensen 1990; Feveile and Jensen 2000; Feveile 2012; Sindbæk 2022a). From these 8th-century origins, Ribe was a significant maritime emporium, part of a wider northern European network of such emporia sites, that emerged even earlier than the well-known sites of Birka, Kaupang or Hedeby (Feveile 2012; Sindbæk 2022b). With easy access to the North Sea, Ribe remained a major hub until at least the latter half of the 9th century, with strong evidence for long distance trade connections with the Continent and as far as the Middle East, as well as specialised craft production in high quality glass, antler and, of course, nonferrous metal goods (Sindbæk 2022c, 435–63).

Ironically, the type of mould used by the metalcasters of Ribe can be best defined as *single-use* investment moulds, made of layered clay and organic material like horsehair (Arrhenius 1973; Brinch Madsen 1984, 82–93; Lønborg and Liversage 1991; Lønborg 1998, 1999; Hedegaard 1992; Söderberg 2002; 2004; Smith 2005; Deckers *et al.* 2021, 33). As the name implies, this is a mould that would have only been used for a single episode of casting before being discarded. Though there has been plenty of debate over the minutia of the process (see Smith 2005), moulds could broadly be produced in one of two ways, either by directly pressing a solid model into the wet clay to create the impression cavity or through the creation of a master wax model, which could then be used in the lost wax method (Brinch Madsen 1984, 82–93). This latter process would have involved the creation of a master mould from a master wax model, which could then be used to reproduce wax copies of the master from which the casting moulds would be produced (Brinch Madsen 1984, 82–93). Solid models could be made of materials such as lead (Mainman and Rogers 2000, 2569; Feveile and Jensen 2006, 144; Pedersen 2016, 38), copper alloy (Oldeborg 1966, 84; Carlsson 1983, fig 15; Ambrosiani and Eriksen 1997, 29), organic material, such as bone, horn, or antler (Cappelle and Vierck 1971, 51, 54; Lønborg 1998; Hedegaard 2005), and previously cast objects (Jansson 1985; Fuglesang 1987). None of these solid models are particularly common in the archaeological record and only represent a narrow range of objects (see Pedersen 2016, 38). Rather, it is likely that beeswax and the lost wax method represented the more common method of mould creation, particularly when it came to some of the more complex jewellery types like oval brooches (Thunmark-Nylen 1983, 7; Brinch Madsen 1984, 82–93; Lønborg 1998; Smith 2005; Gustafsson 2016; Pedersen 2016, 39). Being a soft and malleable material, wax would have been far easier to carve into a positive impression for producing negative mould impressions than softer metals like lead (Gustafsson 2016, 99). Of course, it is also because of this soft and malleable nature that so few models survive: the nature of the lost wax method meant that wax would melt or evaporate during the casting process. The presence of wax in the archaeological record is thus very scant, and there are only a handful of extant examples from the Viking Age, such as the unworked blocks from Gotland and Scania in Sweden (Gustafsson 2016, 99). No surviving wax models for the purpose of casting are currently known (Smith 2005, 84; Pedersen 2016).

Following the creation of the impression, the wet clay mould would be dried, and then fired, at which point it would be ready for casting (Brinch Madsen 1984, 82–93). It would have comprised two separate halves, one for each side of the object, which would have been sealed together with a single opening, into which the metal could be poured (Brinch Madsen 1984, 82–93). Once the metal had cooled, the mould had to be broken to retrieve the now hardened object, which would then be reworked and finished (Brinch Madsen 1984, 82–93).

Interestingly, only the recycling of previously cast brooches as casting models fits any of the conventional definitions of secondary reuse as laid out by Michael Schiffer (1987). Indeed, all other model forms, be they from organic material, lead, copper alloy, or wax, were created with the primary purpose of serving as models, meaning that they are not a part of any system of secondary reuse *per se*. However, this would seem to be a weakness of how reuse has been defined in the archaeological record rather than being indicative of the absence of circularity in the metal-casting process. To follow a broader contemporary definition of circular economy, the goal of circularity is, of course, the reduction of waste, and this certainly would have been achieved through this practice in the casting workshops of Ribe (Kirchherr *et al.* 2017; Murray *et al.* 2017; Winans *et al.* 2017; Korhonen *et al.* 2018).

Northern Emporium

The purpose of the Northern Emporium project from which the material under study emerged was to attain a greater understanding of the production at and long-distance contacts of Ribe during the Viking Age, specifically through the actors and networks active in this early urban environment (Sindbæk 2022b, 18). This was pursued on the basis of a 'high definition' methodology, which involved using a wide variety of scientific methods and micro-approaches, including new radiocarbon and dendrochronological models for absolute dating, 3D spatial recording of stratigraphy, micromorphological analysis of soil, and plant and faunal analysis, to generate a greater wealth of concrete data than had been previously achievable at Ribe (Sindbæk 2022b, 18–19. See also Croix 2022; Jensen 2022; Kveiborg 2022; Phillipsen *et al.* 2022; Wouters 2022, for example).

The assemblage of moulds collected during the Northern Emporium project is staggering, with nearly 7,000 individual fragments emerging from a series of largely 8th- to 9th-century workshops, which produced nonferrous metal jewellery, fittings, keys, coins, amulets, and weapon components (Deckers *et al.* 2021, 31; Croix *et al.* 2022, 49–214). Roughly 600 of these moulds still contain impressions of the objects they were used to cast (Deckers 2021, 31). As an exploratory case study for model tracking, a small sample of six well-defined mould fragments were selected from this larger assemblage. These fragments preserve the impressions of a previously undescribed brooch type that is currently known only from a handful of examples from Denmark and, as such, could be further contextualised with the production data available in the moulds. Only one other star-shaped brooch mould fragment is known from Ribe outside the Northern Emporium excavations, which came from the excavation at Kunstmuseet Have in the 1970s; however, it was not identified at the time, as there were no extant examples (Nancke-Krogh 1978, 180; Brinch Madsen 1984, 156). This type will simply be referred to as the star-shaped brooch type, after their distinct and rather unusual shape (see Feveile 2014, 62).

Star-shaped brooches

There are currently four known extant brooches of the star-shaped type, all from Denmark: two from Funen, in Viby and Faaborg respectively, another from the parish of Hviding, just 8 km south of Ribe, and one from the town of Grenaa, in northeast Jutland (Feveile 2014, 62; Detektor Danmark 2022; Museum of Southwest Jutland 2022; DIME 2023). All four examples are metal-detected finds from plough soil and have very little archaeological context (Feveile 2014, 62; Detektor Danmark 2022; Museum of Southwest Jutland 2022; DIME 2023). Two of these, the Faaborg and Hviding brooches (Pl. 18), are in a poor state of preservation and cannot reveal the full extent of the visual elements that were present (Detektor Danmark 2022; Museum of Southwest Jutland 2022). This is not the case with the examples from Viby (Pl. 19) and Grenaa (Pl. 20), however, which preserve in detail a complex set of artistic motifs and morphological features (Feveile 2014, 62; DIME 2023).

The star-shaped brooch can be defined as an open-work brooch, in the shape of an eight-pointed star with two layers: the underlying first layer consists of an openwork rectangle with the famous gripping beast motif of the Viking Age at the centre, rendered in three dimensions and gripping both sides of the rectangle; the second, or top, layer is an enclosed, open-work cruciform frame, with each point of the cross terminating in an outward-facing beast head with small eyes and a long snout. This layer overlays the first layer in such a way that the points of both shapes do not overlap, thus giving the impression of a star. The stylistic appearance of the central gripping beast allows for an approximate dating in the first half of the 9th century, based largely on contemporary metalwork, such as Jan Petersen's Berdal group of oval brooches (Petersen 1928; Sindbæk 2012).

The mould fragments: establishing a chronology

The precise stratigraphic context in which the moulds from the Northern Emporium excavation originate can further detail their chronological range. There were 18 site phases (with corresponding 'F' numbers) defined in the chronology of the excavation, ranging from the first initial traces of human activity in AD 690 in F1 and F2, through to 2017 in F18 (Croix *et al.* 2022, 55–57). The three most securely dated fragments come from successive building constructions: one (×0427.09) from F11, *c.*AD 810–830, in building K23, and two (×0432.03 and ×411.01) from F12, *c.* AD 830–860, in building K24 (Croix *et al.* 2022, 174, 182). The remaining three fragments can be regarded as intrusive or redeposited finds: ×0017.03, which was found in F18, *c.* AD 1500–2017; ×1056.11, which emerged from the so-called 'dark earth' layers of F9 (*c.* AD 750–790); and ×1481.1, which came out of the stratigraphic profile between F11 and F12 or F13 (*c.* AD 860–880) (Croix *et al.* 2022, 160, 192). This gives an initial range of roughly 120 years, disregarding the obvious redeposition of ×0017.03 in a non-Viking Age layer, though this can be further narrowed by considering the context of the earliest and latest possible phases (F9 and F13). In F9, the evidence for nonferrous metalworking occurs largely near the end of the phase in the form of 155 crucible fragments and 141 casting mould fragments (Croix *et al.* 2022, 156). Though the stratigraphic integrity and chronology of F9 can broadly be considered with a fair degree of confidence, the particular dark earth layers from which fragment ×1056.11 emerges (A343 and A364) are less certain, as they largely consist of poorly separated soil layers from the plot margin and road areas with little in the way of discernible features (Croix *et al.* 2022, 145–146; Deckers 2022, 278). This is further compounded by the fact that no mould fragments for star-shaped brooches occur in F10. While this could be a preservation issue, the uncertain stratigraphy of F9's dark earth layers and the much more secure dating of the fragments in F11 and F12 would indicate that ×1056.11 is an intrusive find in F9. Likewise, the upper limit of the star-shaped moulds, F13, should also be considered with a fair degree of caution, as this phase also saw an increase in displaced finds from younger layers (Deckers 2022, 278). Based solely on the stratigraphy, it is perhaps best to establish a firmer origin date at the start of F11 (AD 810–830) (Croix *et al.* 2022, 174). However, the later chronology is somewhat in doubt, due to the two displaced fragments, most notably ×1481.1, being a loose find between F10 and F13, a later Viking Age phase than the rest of the assemblage. This will be further elaborated below.

Method

To assess for individual models within the type, the mould impressions must be reconstructed and inverted to be properly compared with one another, as well as the extant brooches. Due to their preservation state, this can be a difficult process. Previous methods applied to their study relied solely on autoptic methods of reconstruction, with varying degrees of success and efficiency. The first reconstruction was attempted by Søren Nancke-Krogh in 1978 using the fragments of a single mould for a horse-shaped ornament from the 1970–76 excavations of Ribe. This study was rather novel, utilising photography with light directed from eight different angles to produce a detailed set of photographs which were then used to make a reconstruction drawing of the entire object to fill in the gaps (Nancke-Krogh 1978, 179–182). It demonstrated the importance of technology in approaching such difficult-to-discern production materials and marked a renewed interest in a type of production waste that had not garnered much attention in previous decades, aside from a few works (see Oldeborg 1966, for example). The downside, of course, was that this was a very time-consuming and involved process for the reconstruction of just a single mould and would not be feasible for dealing with any sort of larger assemblage. Furthermore, being limited by the technology of the time, it still ultimately amounts to a purely autoptic approach with a heavy reliance on reconstruction drawings, in addition to reducing comparison to only two dimensions.

Drawings would prove to be the way of the future for mould reconstruction, with two further landmark studies: Helge Brinch Madsen's (1984) study of the 1970s mould corpus and Claus Feveile's (2002) study of the oval brooch mould fragments from Ribe. Both studies left photography behind, opting to rely solely on reconstruction drawings that were often based on surviving brooches rather than the mould fragments for typological and chronological analysis. This allowed for much more efficient study, and the works were able to reconstruct large assemblages rather than just single examples. However, while these works were extremely effective in demonstrating the array of object types being produced, the methods being used to produce them and their potential geographical and chronological range, the autoptic nature of drawing meant that there were almost certainly details lost from the original object that could be vital in tracing an individual model. This loss of detail is well demonstrated in the drawing of the mould fragment from the Kunstmuseet Have excavations (D10448) published by Brinch Madsen (1984, 156), which is quite skewed in the details of the animal head terminals and triangular star point. The result of this skewing is that the impression went unidentified even after the excavation of extant examples for comparison.

3D scanning

The past decade or so has seen the development of a new reconstruction method in the form of 3D scanning (Gustafsson 2011; Neiß *et al.* 2016; Croix *et al.* 2019; Croix *et al.* 2020; Deckers *et al.* 2021). The earliest case study was a small-scale analysis of a poorly preserved mould fragment for a bridle mount from the 11th-century layers of Sigtuna, Sweden. In this study, the fragment was scanned, and the negative impression was inverted in a graphics editor (Gustafsson 2011). The method was demonstrated to be effective at enhancing the decayed and scorched surfaces of the mould, making it an ideal tool for analysis (Gustafsson 2011, 7). Since 2011, of course, the technology has only improved, and a series of investigations into a sample of the Ribe assemblages have shown the technique to be effective in fragment refitting, analysis of production techniques and motif reconstruction (Croix *et al.* 2019; Croix *et al.* 2020; Deckers *et al.* 2021). Sarah Croix *et al.* (2019) holds particular significance here as the first study to specifically use 3D object *comparison*, as opposed to just 3D reconstruction to produce enhanced images, in order to assess the model-identity of individual mould fragments.

For the reconstruction of the Northern Emporium samples, the 3D method was again used, this time with a structured blue light scanner: the AutoScan-DS-MIX, produced by Shining 3D. This scanner can create highly detailed scans of objects up to approximately 10 cm, which can then be converted into 3D digital models using the built-in Ultrascan software. It possesses an accuracy of $\leq$10μm, or 10 microns, comparable to specifications currently used for high-precision dental work and industrial quality inspection (Shining 3D 2022). A single scan takes only 13 seconds, enabling highly efficient processing of larger assemblages if so needed. Once models of the six moulds were rendered with this machine, they were processed in the 3D mesh editing software MeshLab (Cignoni *et al.* 2008, 129–136). This involved colouring and shading the model to provide optimal conditions to view the impression. Cutaways of the digital models were produced so that the inverted positive image could be seen and the object more easily visualised.

After the scans had been properly rendered and edited, they could be manipulated and overlaid with extant brooches in MeshLab to further reduce the margin of error in finding feature matches and differences (Cignoni *et al.* 2008, 129–136). This was rather easily done with the mould fragments, which were all made available to the author for scanning. However, to make a full model comparison of the corpus, it would have also been ideal to overlay the mould fragments with all the extant brooches. Unfortunately, only two of these, the Grenaa and Viby brooches (stored at the Viborg Museum and National Museum of Denmark respectively, DIME ID 198107 and ID C41222), were available for scanning. Comparative analysis with the Faaborg and Hviding brooches was instead performed through autoptic comparison between scans and photos of the extant objects. While still relying on the naked eye to a degree, this method nevertheless had the advantage over drawings or photographs in that the impression reconstruction itself is still produced by the scanner.

Results

Four of the six star-shaped moulds preserve only a small portion of the original impression, typically just including a single animal-head terminal of the upper cruciform layer and occasionally a corner of the lower layer rectangle. These terminals tend to be fairly uniform across all of the impressions, making individual model identification challenging, even with the aid of 3D scanning. Two fragments, however (×432.03 and ×1481.1), preserve a significantly larger portion of the impression, critically containing the head and a single limb of the central gripping beast motif (Pls. 21 and 22, respectively). The 3D scans of these two fragments revealed several key differences in the morphologies of the two beasts. The most immediately apparent of these differences is a lack of elongated ears on the beast of fragment ×1481.1, which fragment ×432.03 possesses in the form of two long, rounded structures directly on top of the head. Furthermore, while both have a rather wide and bulbous cranium, the eyes and snout of the ×432.03 beast are far more elongated than those of ×1481.1, which are more angular. These differences indicate the use of at least two separate models for production. Given the uniformity of the animal-head terminals between impressions, it is possible that these two fragments emerged from the modification of two different wax casts from the same model. Overlay of the Grenaa brooch revealed a strong correlation with ×1481.1, particularly in the angular shape of the snout, as well as the angle of the 'arms' of the star (see Pls. 20 and 22). A third model was revealed when these two moulds were overlaid with a scan of the Viby brooch, which was shown to have rather more elongated animal head terminals, as well as a less clearly defined face on the central gripping beast in which the eyes and ears are simply incised grooves (Pl. 19). This was quite unlike the two Ribe moulds and the Grenaa brooch.

In autoptic comparison between the two extant brooches that were not scanned (Faaborg and Hviding) and the Ribe moulds, it was found that far more similarity existed between the terminal heads than with the brooch from Viby, which could indicate possible matches (see Pl. 17). These terminals are more squat, rectangular and uniform in shape than those of the brooch from Viby. However, the surface corrosion on these two examples made further correlations more difficult to assess, especially when it came to comparing the central gripping beasts. Scanning and overlay of both brooches will be needed before any of these matches can be definitively assessed as corresponding or different models.

Discussion

While there is no strongly corresponding mould that matches the Viby brooch to further clarify the chronology of all three models, this can be achieved with the more securely dated Ribe models. The model impression present in fragment ×1481.1, which was a loose find between F11 and F13, might best be assigned (though not definitively) to F11, given the presence of a different model in a well-established workshop in F12 and the less certain stratigraphy of the later phase, F13. The potential Christian iconography of the brooch also points to a phase earlier than F13. This iconography was certainly not uncommon in the late 8th and early 9th centuries, with cruciform brooches and oval brooches with overt Christian symbols also present both in the mould corpus of Ribe and extant jewellery from around Scandinavia, but it largely disappeared in the jewellery of the latter half of the 9th century (Brinch Madsen 1984; Feveile 2002; 2011; Sindbæk 2014).Taking this into account, F12 (AD 830–860) should be considered the upper limit for this type. Furthermore, as stated above, F11 should likely be considered the beginning of production. The final decade of F9 (AD 780–790) is certainly an intriguing possibility to consider for the beginning of production, as it is in many ways a formative period in the history of metalworking at Ribe: the evidence for metalworking here is sparse, and many designs are unique, represented in only a single mould fragment (Feveile 2002; Croix 2022 *et al.*, 156). Given the rarity of the star-shaped type, it is tempting to say that it might have emerged out of such an experimental phase. However, the issues with the uncertain stratigraphy of the F9 dark earth units and the lack of star-shaped moulds in the stratigraphy in F10 cannot be ignored. This gives an approximate production range of AD 810 to 860, at least at Ribe. With its heavy correlation with the Ribe models, especially ×1481.1, the Grenaa brooch can also be said to fall into this range.

Considering the small sample size of star-shaped brooch impressions in the moulds and the even smaller corpus of extant brooches, the number of individually distinct models that can be discerned with 3D reconstruction is remarkable. What is particularly striking is how similar the two Ribe models are, only varying in the morphology of the head of the central beast. There clearly was an attempt to maintain some level of consistency between models, though why this was the case is more difficult to answer. There are many possibilities here, ranging from a lost or damaged model that resulted in an approximation of the original to be produced, to the hand of a different craftsman putting their own 'stamp' on a familiar design. What is clear, based on the small quantity of these moulds recovered, is that these items were not churned out in the same numbers as certain other types, such as oval brooches of the Berdal type (Petersen 1928, 12; Brinch Madsen 1984; Feveile 2002). Despite this, the presence of different models indicates that even such a rare type cannot be expected to simply correspond to copies from a single master model. It is also possible that there are even fewer moulds here than the fragments would suggest: two fragments (×0432.03 and ×411.01), after all, emerge from the same phase (F12) and could very well be a part of the same mould. Though it has not been possible to determine from these two examples, Croix *et al.* (2020) have demonstrated that mould fragments can be refitted. Furthermore, the low number of fragments over both phases of production, in combination with the rarity of extant examples, indicates that perhaps this was an object that was only ever produced sporadically, on a direct-to-customer commission basis. Such systems have certainly been attested in the Viking Age, for instance on Gotland (Gustafsson 2013, 103). The range of the extant examples is also quite telling; with two from Funen and two from Jutland, this was a localised design in Denmark, and it is likely that demand was simply not high.

Based on the above, it is perhaps not so unusual that an object that occupies such a low percentage of the mould assemblage has a comparatively high percentage of models. Spread over a period of at least 50 years, it makes sense that models would not be retained, especially since this period is likely spread over several generations of craftsmen. This situation would likely be far different if one were to look at the Berdal brooches, for instance, which were much more consistently produced from phase to phase and follow a number of serialised styles (Petersen 1928; Brinch Madsen 1984; Feveile 2002).

Conclusion

Circular processes in the past were prevalent and diverse (Bavuso *et al.* 2023, 3). The Viking Age in Scandinavia was no exception, and many examples of circular practice can be highlighted. What is needed is more lab-based analysis to understand those processes that are not immediately visible in the archaeological record to establish a more holistic and clear view of Viking Age circularity (Bavuso *et al.* 2023, 12). Even a small-scale study on a single craft industry such as this one illustrates the need for new technical approaches to difficult archaeological materials and underlying systemic processes of reuse. In this case, the utility of new methods of 3D visualisation has been demonstrated on a small sample of ceramic casting mould fragments, but the potential here is much greater, particularly with the speed at which modern scanners can process these artefacts. Being able to scan a high volume of objects in a relatively short time span raises the potential of bulk studies of these materials. Furthermore, it is also a point of considerable significance that the moulds and the extant brooches were able to be directly compared in 3D, despite the fact that they are housed in different collections and could not be brought together physically with any convenience. These factors mean that the method has the potential to be quite fruitful with sites like Ribe and Birka, from which thousands of fragments have been excavated (see Ambrosiani 2013; Sindbæk *et al.* 2022), as well as the more heavily deteriorated assemblages of Kaupang (Pedersen 2016, 95). While a bulk study of even one of these sites would provide a wealth of new data on object development, chronologies and production methods, producing 3D datasets from multiple emporia sites could allow for the reconstruction of entire networks of production and should be a priority for future research.

Acknowledgements

This work was supported by the Danish National Research Foundation under the grant DNRF119 - Centre of Excellence for Urban Network Evolutions (UrbNet).

Bibliography

Ambrosiani, B. (2013) *Stratigraphy: Vol. 1, P. 1: The Site and the Shore; P. 2: The Bronze Caster's Workshop.* Stockholm, Riksantikvarieämbetet.

Ambrosiani, B. and Erikson, B.G. (1997) *Birka. Vikingastaden, 5: Vikingastaden lever upp igjen i TV: s modell av 800-talets Birka.* Stockholm, Sveriges Radios Forlag.

Andrén, A. (2013) Places, monuments, and objects: the past in ancient Scandinavia. *Scandinavian Studies* 85.3, 267–281.

Arrhenius, B. (1973) Gjutformar och deglar, påträffade i Birka. In B. Ambrosiani, B. Arrhenius, K. Danielsson, O. Kyhlberg and G. Werner (eds) *Birka. Svarta Jordens Hamnområde. Arkeologisk Undersökning 1970–71*, Riksantikvarieämbetets Rapport C1, 99–114. Stockholm, Riksantikvarieämbetet.

Barfod, G.H., Feveile, C. and Sindbæk, S.M. (2022) Splinters to splendours: from upcycled glass to Viking beads at Ribe, Denmark. *Archaeological and Anthropological Sciences* 14.9, 180.

Bavuso, I., Furlan, G., Intagliata, E.E. and Steding, J. (2023) Circular economy in Roman and early medieval Europe – methods of analysis for a future agenda. *Open Archaeology* 9(1), 20220301. https://doi.org/10.1515/opar-2022-0301

Bencard, M. and Jørgensen, L.B. (1990) The foundation of Ribe. *Antiquity* 64.244, 576–583.

Brinch Madsen, H. (1984) Metal-casting, techniques, production and workshops. In M. Bencard (ed.) *Ribe Excavations 1970–1976* Vol. 2, 15–189. Esbjerg, Sydjysk Universitetsforlag.

Capelle, T. and Vierck, H. (1971) Modeln der Merowinger-und Wikingerzeit. In K. Hauck (ed.) *Band 5 1971*, 42–100. Berlin/Boston, De Gruyter.

Carlsson, A. (1983) *Djurhuvudformiga spännen och gotländsk vikingatid*. Unpublished thesis, Stockholm University.

Cignoni, P., Callieri, M., Corsini, M., Dellepiane, M., Ganovelli, F. and Ranzuglia, G. (2008) MeshLab: an Open-Source Mesh Processing Tool. In V. Scarano, R. De Chiara and U. Erra (eds) *Sixth Eurographics Italian Chapter Conference: Salerno, Italy July 2nd–4th, 2008*, 129–136. Aire-la-Ville, Eurographics Association.

Croix, S. (2022) 3D spatial recording and stratigraphic analysis. In S. Sindbæk (ed.) *Northern Emporium* Vol. 1, 219–238. Højbjerg, Jutland Archaeological Society.

Croix, S., Neiβ, M. and Sindbæk, S.M. (2019) The réseau opératoire of Urbanization: Craft Collaborations and Organization in an Early Medieval Workshop in Ribe, Denmark. *Cambridge Archaeological Journal* 29.2, 345–364.

Croix, S., Deckers, P. and Sindbæk, S.M. (2020) Recasting a Viking warrior woman from Ribe: 3D digital image reconstruction compared. *Journal of Archaeological Science: Reports* 32, 102455.

Croix, S., Deckers, P., Feveile, C., Knudsen, M., Qvistgaard, S., Sindbæk, S. and Wouters, B. (2022) Excavation Atlas. In S. Sindbæk (ed.) *Northern Emporium* Vol. 1, 49–217. Højbjerg, Jutland Archaeological Society.

Deckers, P. (2022) Site formation and artefact assemblages. In S. Sindbæk (ed.) *Northern Emporium* Vol. 1, 271–300. Højbjerg, Jutland Archaeological Society.

Deckers, P., Croix, S. and Sindbæk, S. (2021) Assembling the full cast: ritual performance, gender transgression and iconographic innovation in Viking-Age Ribe. *Medieval Archaeology* 65.1, 30–65.

Detektor Danmark (2022) *Detektor Danmark*. Facebook. Available at: https://www.facebook.com/groups/329154183827681/ (Accessed: November 2022).

DIME. (2023) *Fibel 198107, Digitale Metaldetektorfund*. Available at: https://www.metaldetektorfund.dk/ny/fund/?dimeid=198107 (Accessed: March 1, 2023).

Drescher, H. (1983) Metallhandwerk des 8.–11. Jahrhunderts in Haithabu auf Grund der Werkstattabfälle. In H. Jahnkuhn, H. Tiefenbach, W. Janssen and R. Schmidt-Wiegand (eds) *Das Handwerk in vor- und frühgeschichtlicher Zeit*, vol. II, 174–191. Göttingen, Vandenhoeck und Ruprecht.

Feveile, C. (2002) Støbning af ovale skålspænder i Ribe: type- og teknikvariation. In M. B. Henriksen (ed.) *Metalhåndværk og Håndværkspladser fra Yngre Germansk Jernalder, Vikingetid og Tidlig Middelalder*, 17–26. Odense, Odense Bys Museer.

Feveile, C. (2011) Korsfibler af Råhedetypen: En upåagtet fibeltype fra ældre vikingetid. *Kuml* 60.60, 143–160.

Feveile, C. (2012) Ribe: emporia and town in the 8th and 9th century. In S. Gelichi and R. Hodges (eds) *From One Sea to Another. Trading Places in the European and Mediterranean Early Middle Ages*, 111–122. Turnhout, Brepols.

Feveile, C. (2014) På sporet: af den nordøstfynske vikingetid. Fagarkæologens beretning fra efterårets store detektortræf på Hindsholm. *Fynboer og Arkaeologi* 1, 8–10.

Feveile, C. and Jensen, S. (2000) Ribe in the 8th and 9th century: a contribution to the archaeological chronology of northwestern Europe. *Acta Archaeologica* 71.1, 9–24.

Feveile, C. and Jensen, S. (2006) ASR Posthuset. In C. Feveile (ed.) *Ribe Studier Det ældste Ribe. Udgravninger på nordsiden af Ribe Å 1984-2000* vol 1.2, 119–189. Højbjerg, Jysk Arkæologisk Selskab.

Fuglesang, S.H. (1987) 'The personal touch': on the identification of workshops. In J. Knirk (ed.) *Proceedings of the 10th Viking Congress, Larkollen, Norway, 1985*, 219–230. Oslo, Universitetes Oldsaksamlings.

Gustafsson, N.B. (2011) New light on tempered clay: 3D white light scanning as a means of analysis of early medieval casting moulds. *Lund Archaeological Review* 17, 5–10.

Gustafsson, N.B. (2013) *Casting identities in central seclusion. Aspects of non-ferrous metalworking and society on Gotland in the early medieval period*, PhD Thesis. Stockholm, Stockholm University.

Gustafsson, N.B. (2016) Beeswax in metalworking in Viking period Gotland. *Fornvännen* 111.2, 97–101.

Halperin, C.T. (2021) Ancient recycling: considerations of the wasteful, meaningful, and practical from the Maya site of Ucanal, Peten, Guatemala. *Journal of Archaeological Method and Theory* 27, 766–792.

Hedegaard, K.R. (1992) Bronzestøberhåndværket i yngre germanertid og tidlig vikingetid i Skandinavien–teknologi og organisation. *Lag 1992*, 75–92.

Hedegaard, K.R. (2005) Casting trefoil brooches. *Viking Heritage Magazine* 1.5, 8–13.s

Jansson, I. (1985) *Ovala spännbucklor: en studie av vikingatida standardsmycken med utgångspunkt från Björkö-fynden*. Institutet för Arkeologi Gustavianum, Uppsala University.

Jensen, P.M. (2022) The analysis of eight- and ninth-century plant material from Posthustorvet. In S. Sindbæk (ed.) *Northern Emporium* Vol. 1, 417–433). Højbjerg, Jutland Archaeological Society.

Kirchherr, J., Reike, D. and Hekker, M. (2017) Conceptualizing the circular economy: an analysis of 114 definitions. *Resources, Conservation and Recycling* 127, 221–232.

Klevnäs, A. (2016) 'Imbued with the essence of the owner': personhood and possessions in the reopening and reworking of Viking-Age burials. *European Journal of Archaeology* 19.3, 456–476.

Korhonen, J., Honkasalo, A. and Seppälä, J. (2018) Circular economy: the concept and its limitations. *Ecological Economics* 143, 37–46.

Krogh, S.N. (1978) Ribehesten og dens slægtninge. *Kuml* 27.27, 179–191.

Kveiborg, J. (2022) Zooarchaeology. In S. Sindbæk (ed.) *Northern Emporium* Vol. 1. Højbjerg, Jutland Archaeological Society, 363–388.

Lønborg, B. (1998) *Vikingetidens Metalbearbejdning*. Odense, Odense Bys Museer.

Lønborg, B. (1999) Vikingetidens kvindedragter. *Kuml* 13.13, 259–268.

Lønborg, B. and Liversage, D. (1991) Fremstillingen af vikingetidens skålformede fibler. *Kuml* 38(38), 151–164.

Lund, J. (2022) Kerbing relations through time: Reuse, connectivity and folded time in the Viking Age. *Cambridge Archaeological Journal* 32.2, 245–264.

Lund, J. and Arwill-Nordbladh, E. (2016) Divergent ways of relating to the past in the Viking Age. *European Journal of Archaeology* 19.3, 415–438.

Mainman, A.J. and Rogers, N.S. (2000) *Craft, Industry and Everyday Life: Finds from Anglo-Scandinavian York*. York, Council for British Archaeology.

Maixner, B. (2005) Die gegossenen kleeblattförmigen Fibeln der Wikingerzeit aus Skandinavien. *Germania: Anzeiger der Römisch-Germanischen Kommission des Deutschen Archäologischen Instituts* 85.2, 475–478.

Müller-Wille, M. (2003) The cross goes north: Carolingian times between Rhine and Elbe. In M. Carver (ed.) *The Cross Goes North: Processes of Conversion in Northern Europe, AD 300–1300*, 443–462. Woodbridge, York Medieval Press.

Murray, A., Skene, K. and Haynes, K. (2017) The circular economy: an interdisciplinary exploration of the concept and application in a global context. *Journal of Business Ethics* 140.3, 369–380.

Museum of Southwest Jutland (2022) *Samlingen Online*. http://Sol.Sydvestjyskemuseer.Dk/?Filter=0&search=.

Nancke-Krogh, S. (1978) Ribehesten og dens slægtninge. *Kuml* 27(27), 179–191.

Neiß, M., Sholts, S.B. and Wärmländer, S.K. (2016) New applications of 3D modeling in artefact analysis: three case studies of Viking Age brooches. *Archaeological and Anthropological Sciences* 8, 651–662.

Oldeberg, A. (1966) *Metallteknik under vikingatid och medeltid*. Stockholm, Seelig.

Orfanou, V., Birch, T., Sindbæk, S.M., Feveile, C., Barfod, G.H. and Lesher, C.E. (2021) On diverse arts: crucible metallurgy and the polymetallic cycle at Scandinavia's earliest Viking town, Ribe (8th–9th c. CE), Denmark. *Archaeological and Anthropological Sciences* 13, 1–23.

Parrott, D. (2020) Gotland's box brooches: Using object biography to answer questions of ethnicity. *Apardjón Journal for Scandinavian Studies* 1, 87–111.

Pedersen, A. (2006) Ancient mounds for new graves. In A. Andrén, K. Jennbert and C. Raudvere (eds) *Old Norse Religion in Long-term Perspectives. Origins, Changes and Interactions*, 346–353. Lund, Nordic Academic Press.

Pedersen, U. (2016) *Into the Melting Pot: Non-Ferrous Metalworkers in Viking-Period* Kaupang. Aarhus, Aarhus University Press.

Pedersen, U. (2017) Viking-period non-ferrous metalworking and urban commodity production. In Z. Glørstad and K. Loftsgarden (eds) *Viking-Age Transformations: Trade, Craft and Resources in Western Scandinavia*, 124–138. London, Routledge.

Peña, J.T. (2020) Recycling in the Roman world: concepts, questions, materials, and organization. In C.N. Duckworth and A. Wilson (eds) *Recycling and Reuse in the Roman Economy*, 9–58. Oxford, Oxford University Press.

Petersen, J. (1928) *Vikingetidens Smykker*. Stavanger, Stavanger Museum

Philippsen, B., Feveile, C., Olsen, J. and Sindbæk, S.M. (2022) Single-year radiocarbon dating anchors Viking Age trade cycles in time. *Nature* 601.7893, 392–396.

Schiffer, M.B. (1987) *Formation Processes of the Archaeological Record*. Salt Lake City, University of Utah Press.

Shining 3D (2022) *AutoScan-DS-MIX*. https://Www.Shining3ddental.Com/Solution/Autoscan-Ds-Mix/.

Sindbæk, S.M. (2012) Enter the gripping beast: innovations and actor-networks in Viking Age towns. In R. Berge, M. Jasinski and K. Sognnes (eds) *N-TAG TEN: Proceedings of the 10th Nordic TAG Conference at Stiklestad, Norway 2009*, 129–135. Oxford, British Archaeological Reports.

Sindbæk, S.M. (2014) Crossbreeding beasts: Christian and non-Christian imagery in oval brooches. In I. Garipzanov (ed.) *Conversion and Identity in the Viking Age*, 167–193. Turnhout, Brepols.

Sindbæk, S.M. (ed.) (2022a) *Northern Emporium* Vol. 1. Højbjerg, Jutland Archaeological Society.

Sindbæk, S.M. (2022b) Emporia, high-definition archaeology, and Ribe. In S. Sindbæk (ed.) *Northern Emporium* Vol. 1, 11–23. Højbjerg, Jutland Archaeological Society.

Sindbæk, S.M. (2022c) The making of a northern emporium. In S. Sindbæk (ed.) *Northern Emporium* Vol. 1, 435–463. Højbjerg, Jutland Archaeological Society.

Smith, M.H. (2005) Breaking the mould: A re-evaluation of Viking Age mould-making techniques for oval brooches. In R. Bork, S. Montgomery, C. Neuman de Vegvar, E. Shortell and S. Walton (eds) *De Re Metallica*, 81–99. Aldershot, Ashgate Publishing.

Söderberg, A. (2002) Metalliska spår efter gjuteriverksamhet: en skiss till en arkeologisk fältmetod. *Fornvännen* 97.4, 255–264.

Söderberg, A. (2004) Metallurgic ceramics as a key to Viking Age workshop organisation. *Journal of Nordic Archaeological Science 14*, 115–124.

Thumark-Nylén, L. (1983) Gotland och Baltikum. In I. Jansson (ed.) *Gutar och vikingar*, 306–322. Stockholm, Statens Historiska Museum.

Winans, K., Kendall, A. and Deng, H. (2017) The history and current applications of the circular economy concept. *Renewable and Sustainable Energy Reviews* 86, 825–833.

Wouters, B. (2022) Soil and sediment micromorphology. In S. Sindbæk (ed.) *Northern Emporium Vol. 1*, 239–270. Højbjerg, Jutland Archaeological Society.

Chapter 10

Seeking the invisible with legacy data: notes on the use of archives for the study of ancient circular economies

Emanuele E. Intagliata

Introduction

The study of the circular economy in past societies has seen a surge of interest in the past decades. Regenerative and conservative processes such as recycling, reuse and repair (the so-called 'three Rs') have been the subject of increasingly targeted studies within new research projects and fieldwork. This growing interest has brought about important changes in the way ancient economies are perceived. For example, if the study of economic circularity was once traditionally mostly focused on the late antique and early medieval periods, recently there have been increasing attempts to go beyond these chronological boundaries (*e.g.*, Peña 2007; Duckworth and Wilson 2020).

Despite this growing interest, the approach of the scholarly community to the study of economic circularity in past societies remains patchy. Studies dedicated to the topic often fail to address the totality of circular economic practices with the result that some processes (*e.g.*, reuse) and materials (*e.g.*, reused architecture – so-called *spolia*) have been better studied than others (*e.g.*, repair, see Jervis and Kyle 2012). Archaeometric analyses have oftentimes exacerbated this tendency by exploring the topic using a case-study approach and narrow chronological boundaries. This has created an incomplete understanding of economic circularity that fails to approach this topic from a wider chronological and geographical perspective. Recent scholarship (Bavuso *et al.* 2023) has advocated the need to go beyond this research sectorialisation and explore the circular economy holistically and with a *longue-durée* perspective (for the meaning of this term, see esp. Braudel 1958). This also entails addressing important research gaps through the examination of unpublished data, such as those housed in archaeological archives.

In some cases, archives can represent the only way to gather information on archaeological records that are impossible to examine autoptically, such as in the case of sites in war-torn countries or excavated monuments that have been dismantled for modern construction. In these archives, neglected data can exist that shed light on processes of economic circularity originally ignored by the excavators. Can we go back to legacy data to explore past circular economies and fill in the gap left by past fieldwork? What are the limits of this approach? This paper aims to raise awareness about the potential of legacy data to study circular economic practices based on two case studies in Europe, namely the Archive Collart at the University of Lausanne and the Dörner Archive at the Forschungsstelle Asia Minor, University of Münster. It will demonstrate that, although it is possible to fill in lacunae left by past excavations, there exist limitations in the study of datasets retrieved from archives. This can make it difficult to understand these phenomena thoroughly without examining the physical archaeological evidence.

Understanding data collection

The study of legacy data to approach circular economic processes is not without issues. Although archaeologists strive to be as objective as possible in the field, data collection can be dramatically affected directly or indirectly by the cultural background, research agendas and technological capabilities of an archaeological team – all elements that have collectively been referred to as 'scaffolding' in modern literature (Wylie 2017). Data on circular economic processes were particularly prone to be ignored or overlooked in past excavations, therefore leading to important lacunae in the archaeological record that are now difficult to overcome. The excavations of the Sanctuary of Baalshamin at Palmyra can be taken as an exemplary case study to demonstrate the potential and problems in working with archive datasets to shed light on the reuse of building material.

The Sanctuary of Baalshamin was excavated by a Swiss team led by Paul Collart between 1954 and 1966 (on the history and topography of the monument, see especially Collart and Vicari 1969). The original documents produced during this excavation are now housed at the Archive Collart of the University of Lausanne in Switzerland (Michel 2008). The sanctuary, which underwent major building transformations, particularly in the 1st and 2nd centuries AD, centres around a prostyle temple (Fig. 10.1) that was flanked by two courtyards, namely the *Cour Sud* and the *Grande Cour* or *Cour Nord*. After the 3rd century AD and throughout Late Antiquity and the early Islamic period, the sanctuary was repurposed; according to the excavators, the temple was transformed into a church and the northern courtyard was occupied by other structures, most likely private residential buildings. Recent studies based on a reassessment of the documentation housed in Lausanne and the published evidence have been able to add more to Collart's conclusions by clarifying the gradual process of development of the buildings in the *Grande Cour* and its chronology (Intagliata 2016). In addition, Collart's theory about the transformation of the temple into a church has been put in doubt (Kowalski 1996).

Reused architectural and sculptural material was employed abundantly in the late antique and early Islamic structures built on top of the earlier sanctuary (Fig. 10.2). The practice of reuse was well-known to Collart and his team, who in several cases tried to delve into its significance and meaning. For example, the reuse of sculpture and architecture from the pagan sanctuary in the back wall of the alleged church was interpreted as a message epitomising the triumph of Christianity over paganism (Collart and Vicari 1969, 81). Such a traditional approach to the meaning of reuse was not new and was in line with that of other archaeological teams operating in Palmyra at that same time (see Intagliata 2021 for further discussion).

Fig. 10.1 The remains of the temple of Baalshamin emerging from the sands in 1954 (© UNIL IASA; Patrick Michel).

Fig. 10.2 Examples of the reuse of architectural and sculptural material from the sanctuary of Baalshamin (© UNIL IASA; Patrick Michel).

Despite this, the study of this building practice was not on Collart's list of priorities. This led to gaps in the documentation that are evident when examining the relevant publications and unpublished documents and photographs housed in Lausanne. For example, all information about the location of sculptures in late antique and early Islamic walls can be surmised by reading the notes in the excavation diaries and, therefore, the published catalogue dedicated to this material (Dunant and Stucky 2000). However, in most cases, the position of the material is not clearly indicated, leaving the researcher in doubt as to where and how exactly a sculpture was embedded into a wall. Reused inscriptions are better documented – a published plan shows where epigraphic material was found within the building. Yet, there is no information about where exactly these inscriptions were embedded into a wall or whether the scripts could be visible to passers-by. Therefore, it remains to be understood whether inscriptions on stone supports were reused because of their aesthetic/ideological appeal or were employed as mere building materials. Finally, as opposed to inscribed or decorated architectural elements and sculpture, undecorated architecture was not thoroughly documented. All this leaves us with an important gap to fill, for a thorough catalogue of reused material cannot be written with the data at hand.

Photographs that document the existence of reused inscriptions and sculptures exist, but they are problematic in many ways. Most of the reused material was photographed only after it was removed from walls and in a neutral background – often on a piece of textile or on the ground. This is indicative of how Collart and his team approached the study of this monumental complex during their excavations. In photographing the reused artefact detached from its late antique and early Islamic context, the archaeologist decides to focus the attention of the viewer only on one part of the history of the object – its role in the sacred space before the 4th century AD – while neglecting others.

Fig. 10.3 Excavation of the Grande Cour *of the Sanctuary of Baalshamin. Note the aligned reused column drums in the background (© UNIL IASA; Patrick Michel).*

In most cases, our understanding of how reused materials were embedded into walls comes from photographs that were not meant to document artefacts *in situ* but rather to document the progressive state of the excavations in the sanctuary. Therefore, a photograph of the excavation of the northern courtyard of the sanctuary may unintentionally contribute to pinpointing the location of reused material and how these were used as building blocks (Fig. 10.3). Cases in which such material is the subject of photographs are rare. These include mostly photographs of architectural elements (especially capitals), while no photographs exist of sculptural elements embedded into walls.

The fact that the data collection stage failed to document systematically the phenomenon of reuse before the dismantling of the late-antique and early Islamic structures within the sanctuary can be explained as the result of a combination of several factors, among which is Collart's research agenda. As Collart and Jacques Vicari openly admit in the first volume of the series dedicated to this compound, the team was first attracted to Palmyra by its monumental remains and history. Collart believed the Sanctuary of Baalshamin to be a particularly appealing target for archaeological investigations due to the role that this deity played in the Palmyrene religion and the monumental remains of the temple that were still visible emerging from the sand in 1954:

> Le choix du site de Palmyre n'a guère besoin d'être justifié: le destin prestigieux de cette ville, la splendeur de ses ruines encore visibles l'avaient désignée d'emblée à notre attention. Et à Palmyre, le sanctuaire de Baalshamîn nous avait aussitôt paru susceptible d'investigations fructueuses, tant en raison du rôle de ce dieu dans la religion des Palmyréniens que de l'ignorance où l'on restait encore des constructions qui devaient exister autour de son temple (Collart and Vicari 1969, 5).

Collart's approach towards the history of Palmyra focused on the study of the pre-4th-century remains. This explains his desire to reconstruct the sanctuary by anastylosis and dismantle all late antique and early Islamic constructions that had occupied the monument:

> Elle [the *Gran Cour*] apparat tout encombrée de murs tardifs de trés mauvaise qualité faits de terre et de blocs de remploi [...]. Le désir de récuperer ceux-ci, plus encore que celui de rétablir dans son intégrité l'état antique du sanctuaire, nous en conseilla la demolition. L'intérèt des documents ainsi retrouvés et les enseignements qu'ils fournirent vinrent justifier pleinement après coup la légitimité de cette suppression (Collart and Vicari 1969, 15–16).

Such a research focus is not peculiar to Collart and his team but fits within a much wider trend in the mid-20th century. Since the first western travels and exploration to Palmyra, the post-Zenobian history of this site had often been considered an uneventful time of decline for the city. This contributed to discouraging systematic studies on this phase of the settlement (for the history of studies on late-antique and early Islamic Palmyra, see Intagliata 2018, ix–xii).

Another factor behind this gap in data is that, although the existence of the practice of reuse of architectural and sculptural elements was already well-known in the 1950s and 1960s, at the time of Collart, a standard framework of reference that would help to approach the study and documentation of this material did not exist. Therefore, simply blaming the excavators for any research gap in the dataset associated with circular economic processes would mean making the mistake of ignoring the 'scaffolding' within which Collart and his team were operating.

In the decades following the publication of the results of the excavations of the Sanctuary of Baalshamin, studies on the reuse of building materials have multiplied. These have contributed to shifting our attention from interpretations associated with the function of each individual reused element to the workforce responsible to build structures with reused material (*e.g.*, Frey 2016). At the site level, studies dedicated to late antique and early Islamic Palmyra have also become more numerous (*e.g.*, in the last decade Genequand 2012, 17–37, 45–67; Intagliata 2018 on the history and archaeology of Palmyra after Zenobia), allowing more light to be shed upon a once-neglected phase of the history of the site. Palmyra is now known to have survived the 3rd-century crisis and to have become an important stronghold along the eastern Roman frontier in Late Antiquity. The site continued to be inhabited after the Islamic takeover in AD 634. By the 9th century, the city had experienced a contraction that gradually moved the settlement into the temenos of the Sanctuary of Bel.

This new conceptual framework, which is the 'scaffolding' of our time, has provided the basis to review the practice of the reuse of architecture and sculpture in the Sanctuary of Baalshamin. Based on the admittedly patchy data from the archive, recent study (Intagliata 2021; 2023) has pinpointed at least two ways in which walls with reused architecture were constructed: one in which the reused material was bulky, heavy, used in all courses of a wall and sometimes retrieved at a considerable distance from the construction site; and one in which reused materials were medium sized and only situated in the lower parts of walls. These two building practices might have been a response to the existence of different types of workforces, namely one that was well organised and large, and one with limited tools for lifting heavy weights and not particularly numerous.

Despite the obvious limitations of the dataset, the identification of these two building patterns contributes to reaching a more nuanced picture of the use of reused architectural and sculptural material in Palmyra. Not all walls built with reused building elements are the same, and the simple exercise of comparing how this material was embedded in a wall provides important insights into the people responsible for its construction. Such a conclusion goes beyond the ideological interpretation followed by Collart and other scholars for Palmyra and opens the possibility of reviewing the phenomenon at the site level. More generally, it demonstrates how ever-changing conceptual frameworks and approaches can transform archives from passive repositories of archaeological documentation into active laboratories for research. This is not to say, however, that archive archaeology can contribute to filling in all lacunae for the study of circular economic processes, for often data can be irremediably lost during the collection stage.

'Invisible' circular economies

The results of the excavation of the so-called 'Iron Field' at Eski Kăhta by Friedrich Karl Dörner and his team are also indicative of the problems that archaeologists can face when working with legacy data to understand circular economies in antiquity. However, in this case, the situation is further complicated by the fact that Dörner might not have been aware that he and his team had retrieved evidence of circular economic practices.

Eski Kăhta (ancient Arsameia-on-the-Nimphayos) is situated in modern Adıyaman province, in the ancient Hellenistic Kingdom of Commagene (Blömer and Winter 2012; Brijder 2014; Blömer *et al.* 2021 on the history of Commagene). The site is best known for being the burial place of King Mithradates I, father of Antiochos I. It is characterised by the presence of two hills, Yeni Kale, occupied by a Mamluk fortress, and Eski Kale. Fieldwork at Eski Kăhta commenced in 1953 and continued not without interruption until 1969. If at the beginning of the excavations the attention of Dörner had focused on the processional road that led to Eski Kale, later works in 1963–1965 and 1967–1969 targeted the hinterland to try to identify evidence of a city reported by the main cult inscription of the *hierothesion* (Dörner and Goell 1963, 41).

Despite all efforts, the 1960s excavations did not yield any evidence for the existence of an urban settlement in the Hellenistic period but brought to light a wealth of data associated with the medieval occupation of the area, particularly at Yeni Kale, Flur Köyönü and the so-called 'Mosaic Field' and 'Iron Field.' A significant portion of the results of the 1960s excavations is still waiting to be fully published. Part of the original documentation of the excavations at Eski Kăhta is now housed at the Forschungsstelle Asia Minor – University of Münster, a research centre founded by the same Dörner to encourage research on the heritage of Turkey (Blömer and Intagliata 2022).

The excavation of the 'Iron Field' was conducted in 1963 and 1965. Dörner and his team were first attracted to this area, which is situated at the foot of Yeni Kale, because of the high concentrations of iron slags. Several trenches with different orientations were laid down to try to cast light on the source(s) of this material. Eventually, one of the trenches intercepted a battery of furnaces for metal production that confirmed the industrial nature of this area of the site. The chronology of the compound remains to be fully disentangled based on a thorough analysis of the archival material. However, a generic medieval chronology seems to be assured based on a cursory cross-examination of unpublished data (especially private correspondence and personal notes taken by Dörner and his team) and published data (Dörner 1965a, 29–31; 1965b; Dörner *et al.* 1965a, 139–145; 1965b, 222–231; 1968, 41–44; see also Intagliata forthcoming).

Dörner was aware of the importance of the excavation of the 'Iron Field' to attract funding from the iron industry in Germany, particularly the *Verein Deutscher Eisenhüttenleute* (VDEh) 'Association of German steel manufacturers' (see Intagliata forthcoming, for a thorough analysis of the history of the excavations and the methodology used by Dörner and his team). This is indicated by the relatively high number of publications describing the work in the 'Iron Field' as opposed to those, for example, for the 'Mosaic Field.' Nonetheless, excavations stopped in 1965 after only two fieldwork seasons.

Dörner's research agenda at Eski Kăhta was undoubtedly directed toward understanding the Hellenistic phase of the settlement. Indeed, the 1960s' expansion of the excavations in the hinterland of this site was meant to intercept the Hellenistic phase of the settlement. Such a research agenda contributed significantly to shaping interpretations in the field. For example, the bathhouse at the 'Mosaic Field' was dated by Dörner and his team to the Hellenistic or early Roman periods, even though the geometric-patterned mosaic found in this compound can be best dated to Late Antiquity (Blömer and Intagliata 2022, with extensive bibliography). Similarly, in several of his publications, Dörner does not exclude the idea that the medieval productive complex at the 'Iron Field' was placed on top of earlier furnaces, even though the evidence from the ground suggested otherwise. Therefore, it is possible that Dörner's research agenda, coupled with disagreement among team members, eventually contributed to bringing the excavations to a halt (Intagliata forthcoming, with extensive bibliography).

The way in which data was collected and stored by Dörner and his team was decentralised. Each trench supervisor and find specialist produced their documentation in the field, and these were not collected by the field director at the end of the excavation but dispersed among team members (Intagliata forthcoming). Despite this, most of the documents associated with the excavation of the 'Iron Field' found their way into the Dörner Archive. At the death of Walter Nowothnig in 1971, who had been trench supervisor during the 1965 season, Dörner retrieved all of the documentation produced during the excavation that then entered the Dörner Archive. These include 59 plans, hundreds of photographs, a list of photographs detailing the exact location of each shot, drawings of artefacts, a list of drawings, manuscripts of publications, some 180 private letters between Dörner and the excavators of the 'Iron Field' and even a short, 20-minute film for the public. Most of this material remains to be sorted, digitised and properly studied (Blömer and Intagliata 2022).

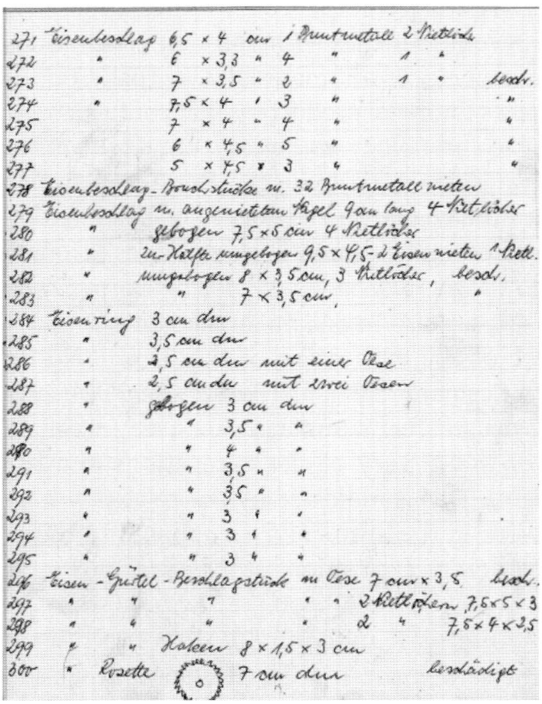
271 Eisenbeschlag 6,5 × 4 cm 1 Buntmetall 2 Nietlöcher
272 " 6 × 3,3 " 4 " 1 "
273 " 7 × 3,5 " 2 " 1 " besch.
274 " 7,5 × 4 " 3 " "
275 7 × 4 " 4 " "
276 6 × 4,5 " 5 " "
277 5 × 4,5 " 3 " "
278 Eisenbeschlag-Bruchstücke m. 32 Buntmetallnieten
279 Eisenbeschlag m. angenietetem Nagel 9 cm lang 4 Nietlöcher
280 " gebogen 7,5 × 5 cm 4 Nietlöcher
281 " zur Hälfte umgebogen 9,5 × 4,5 - 2 Eisennieten 1 Nietl.
282 " umgebogen 8 × 3,5 cm, 3 Nietlöcher, besch.
283 " " 7 × 3,5 cm, "
284 Eisenring 3 cm dm
285 " 3,5 cm dm
286 " 2,5 cm dm mit einer Öse
287 " 2,5 cm dm mit zwei Ösen
288 " gebogen 3 cm dm
289 " " 3,5 " "
290 " " 4 " "
291 " " 3,5 " "
292 " " 3,5 " "
293 " " 3 " "
294 " " 3 " "
295 " " 3 " "
296 Eisen-Gürtel-Beschlagstück m. Öse 7 cm × 3,5 besch.
297 " " " " " 2 Nietlöchern 7,5 × 5 × 3
298 " " " " 2 " 7,5 × 4 × 2,5
299 " " Haken 8 × 1,5 × 3 cm
300 " Rosette 7 cm dm beschädigt

Fig. 10.4 One page from the list of drawings of metal artefacts from the 'Iron Field' (© Forschungsstelle Asia Minor).

The way in which the documentation was managed has caused important lacunae in the archive. For example, correspondence between Dörner and Nowothnig suggests that Erwin Lucius was responsible for studying the pottery from this site and writing a ceramic report. This report is mentioned in several letters but is not present in the archive. Similarly, there exists a document in the Dörner Archive that list all numismatic finds retrieved during the 1965 excavation. However, this list, which may also include material from the 'Iron Field', is incomplete, as detailed by letters written by Dörner to Nowothnig. In these letters, Dörner particularly specifies that data on medieval coins are still missing (Intagliata forthcoming).

The list of drawings of metal finds housed in the archive was written in 1965 – the last year of the expedition in the 'Iron Field' (Fig. 10.4). It includes information on approximately 350 metal artefacts found in the rooms where the metal production facilities were sited. Each artefact is given a progressive number after which follows information on its nature and dimensions; additional notes also provide information about the find (*e.g.*, decoration, number of rivets) and are followed by letters, presumably the initials of the person responsible for the drawing. In a few cases, a sketch of the artefact is also provided. The vast majority of the drawn artefacts (over 280) are classified as 'Eisenbeschlag' (lit. 'iron fitting').

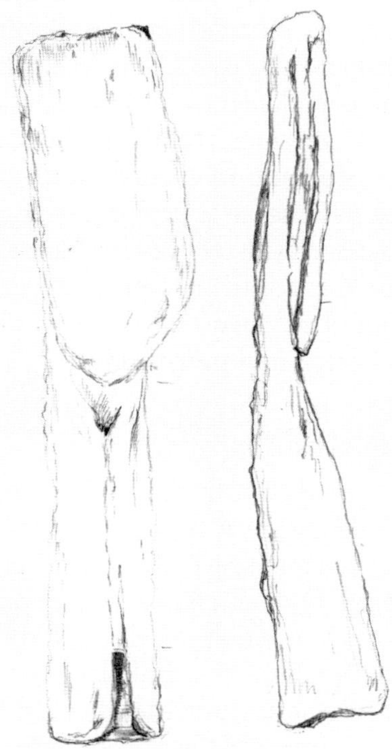

Fig. 10.5 Drawings of a bent spearhead from the 'Iron Field.' Scale unknown (© Forschungsstelle Asia Minor).

Out of the *c.* 350 artefacts in this list, the archive houses the original drawings of slightly over 130 of them. In these drawings, the Eisenbeschlag described in the list of drawings are mostly rendered as metal sheets provided with rivets. Other drawings suggest the high level of variability of the remaining objects. The assemblage included all sorts of weapons – spearheads, arrowheads, halberd heads – everyday objects like keys, belt buckles, knives, a large selection of what appear to be carving tools – point chisels, flat straight chisels, pickaxes – and fragments of metal vessels. Worthy of note is also the wide typological variety of certain artefacts, such as arrowheads, which are well represented in the assemblage and were found in different forms and dimensions.

Future research on the chronology and exact find spot of these artefacts will certainly contribute to casting light on the nature of this assemblage. However, at this early stage of the research, it is already possible to put forward a working hypothesis. The variety of metal objects found in the 'Iron Field' and the fact that several of these types, such as arrowheads and spearheads, were found bent or blunt (Fig. 10.5) may suggest that scavenging activities were taking place at this site. It is possible that several of these objects, which could not be used for their original intended purpose because they were damaged, were collected to be repaired or melted and transformed into something else. Confirming this hypothesis might entail examining the surface of the artefacts themselves and perhaps might involve archaeometallurgists to conduct targeted archaeometric analyses. Any conclusions on the nature of this assemblage would remain hypothetical and uncertain without further examination of the material, whose whereabouts are currently unknown.

It is in this uncertainty that likely lies the future of archives as resources for studying ancient circular economies. Because of their nature, archaeological archives will probably never be able to fill completely the data lacunae left by past excavations. However, they can complement well what new, fresh data from archaeological research have to offer.

Much can certainly still be done to fill the gaps left by previous scholarship through a systematic re-evaluation of old data. Optimism will certainly characterise future studies that attempt to use archives as a tool to understand circular economies in past societies. However, the case studies of the sanctuary of Baalshamin at Palmyra and Eski Kăhta in Commagene presented above have demonstrated that this optimism cannot be open-ended, for it is limited by the nature of the data and how they were collected and processed.

Conclusions

The study of the circular economy has so far been patchy. In general, one can note a tendency towards examining this topic based on targeted case studies that are limited in time and space. Recent literature has advocated the need to approach the study of this phenomenon in antiquity more holistically and with a *longue-durée* perspective. To do so, one would need to fill in the lacunae left by previous fieldwork. Archaeological archives can be useful places in which to look for evidence of economic circularity in past societies and, therefore, to help fill in these lacunae.

The scope of this paper was to raise awareness of the significance of archival studies for shedding light on circular economic processes, such as reuse, recycling and repair, and to highlight several issues associated with this approach. The case studies of the Sanctuary of Baalshamin at Palmyra and Eski Kăhta have demonstrated that it is possible to go back to old data and reach new conclusions on circular economic processes, whether already identified at the time of the excavation or not. However, research agendas, preconceptions and technological capabilities of the excavation teams responsible for data collection at the time when the archive was formed are elements that can contribute to the creation of important limitations in the dataset. In some cases, these limitations are difficult if not impossible to overcome without access to the sites or artefacts in question.

Acknowledgements

Parts of this research were funded by the Danish National Research Foundation under the grant DNRF119 – Centre of Excellence for Urban Network Evolutions (UrbNet). This contribution was also the result of a research trip conducted in November 2021 and funded by the Danish Institute in Damascus.

I am grateful to Patrick Michel, Anne Bielmann (Fonds d'Archives Paul Collart, University of Lausanne) and Michael Blömer (Dörner Archive, University of Münster – Forschungsstelle Asia Minor) for having allowed me to peruse the archives and publish several photographs in this contribution. Any mistakes, however, remain wholly my own.

Bibliography

Bavuso, I., Furlan, G., Intagliata, E.E. and Steding, J. (2023) Circular economy in Roman and early medieval Europe – methods of analysis for a future agenda. *Open Archaeology* 9(1), 20220301. https://doi.org/10.1515/opar-2022-0301

Blömer, M. and Intagliata, E.E. (2022) Archive archaeology at the Forschungsstelle Asia Minor, University of Münster: the Dörner archive and its untapped potential. In O. Bobou, A.C. Miranda and R. Raja (eds) *Archival Historiographies: the impact of 20th-Century Legacy Data on Archaeological Investigations*, 145–159. Turnhout, Brepols.

Blömer, M. and Winter. E. (2012) *Commagene. The Land of Gods between Taurus and Euphrates*. Istanbul, Ege Yayınları.

Blömer, M., Riedel, S., Versluys, M.J. and Winter, E. (eds) (2021) *Common Dwelling Place of All the Gods. Commagene in its Local, Regional, and Global Context* (Oriens et Occidens 34). Stuttgart, Steiner.

Braudel, F. (1958) Histoire et sciences sociales: la longue durée. *Annales. Économies, Sociétés, Civilisations* 13.4, 725–753.

Brijder, H.A.G. (2014) *Nemrud Dağı. Recent Archaeological Research and Conservation Activities in the Tomb Sanctuary on Mount Nemrud*. Berlin, De Gruyter.

Collart, P. and Vicari, J. (1969) *Le Sanctuaire de Baalshamin à Palmyre (I). Topographie et architecture*. Roma, Istituto Svizzero di Roma.

Dörner, F.K. (1965a) Zusammengassender Bericht über die Ausgrabungen in Arsameia 1963. *Türk Arkeoloji Dergisi* 12.2, 24–33.

Dörner, F.K. (1965b) Kommagene. Das Land ubi ferrum nascitur. *Anzeiger der Akademie der Wissenschaften* 102, 1–7.

Dörner, F.K. and Goell, T. (1963) *Arsameia am Nymphaios: die Ausgrabungen im Hierothesion des Mithradates Kallinikos von 1953–1956*. Berlin, Gebr. Mann.

Dörner, F.K., Böhne, K., Brödner, E., Lucius, E. and Nowothnig, W. (1965a) Arsameia am Nymphaios: Bericht über die Grabungskampagne 1965. *Istanbuler Mitteilungen* 16, 130–156.

Dörner, F.K., Hoepfner, W., Müller-Beck, H. and Winkelmann, W. (1965b) Arsameia am Nymphaios. Bericht über die 1963 und 1964 ausgeführten Ausgrabungen, *Archäologische Anzeiger* 2, 189–235.

Dörner, F.K., Böhne, K., Brödner, E., Lucius, E. and Nowothnig, W. (1968) Über Die Ausgrabungen in Arsameia 1965. *Türk Arkeoloji Dergisi* 15.1, 35–57.

Duckworth, C.N. and Wilson, A. (eds) (2020) *Recycling and Reuse in the Roman Economy* (Oxford Studies on the Roman Economy). Oxford, Oxford University Press.

Dunant, C. and Stucky, R.A. (2000) *Le Sanctuaire de Baalshamin à Palmyre (IV). Skulpturen/sculptures*. Roma, Istituto Svizzero Italiano.

Frey, J.M. (2016) *Spolia in Fortifications and the Common Builder in Late Antiquity*. Leiden/Boston, Brill.

Genequand, D. (2012) *Les établissements des élites omeyyades en Palmyène et au Proche-Orient*. Beirut, IFPO.

Intagliata, E.E. (2016) The post-Roman occupation of the northern courtyard of the Sanctuary of Baalshamin in Palmyra; a reassessment of the evidence based on the documents at the Fonds d'archives Paul Collart, Université de Lausanne. *Zeitschrift* für Orient-*Archäologie* 9, 180–199.

Intagliata, E.E. (2018) *Palmyra After Zenobia (ad 272–750)*: An Archaeological and Historical Reappraisal. Oxford, Oxbow Books.

Intagliata, E.E. (2021) Reuse of building material and sculpture in late antique and early Islamic Palmyra (273–750 CE). An overview of the practice and several remarks on the evidence from the Sanctuary of Baalshamin. *Syria* 98, 419–438.

Intagliata, E.E. (2023) A matter of size. A dimensional approach to the study of reused inscriptions and sculptures from the Sanctuary of Baalshamin at Palmyra. In R. Raja and N. Andrade (eds) *Exchange and Reuse: Economy and Circularity at Palmyra*, 99–111. Turnhout, Brepols.

Intagliata, E.E. (forthcoming) Placing the container before the content: the cases of the 'Iron Field' and 'Mosaic Field' at Eski Kăhta at the Dörner Archive – Forschungsstelle Asia Minor, Münster.

In R. Raja (ed.) *Shaping Archaeological Archives: Dialogues between Fieldwork, Museum Collection, and Private Archives*. Turnhout, Brepols.

Jervis, B. and Kyle, A. (eds) (2012) *Make Do and Mend: Archaeologies of Compromise, Repair and Reuse* (BAR International Series 2408). Oxford, British Archaeological Reports.

Kowalski, S.P. (1996) Doubtful Christian reutilization of the Baalshamin temple in Palmyra. *Damaszener Mitteilungen* 9, 217–226.

Michel, P. (2008) Le Fonds d'archives Paul Collart à l'université de Lausanne (Suisse). *Anabases* 7, 241–249.

Peña, J.T. (2007) *Roman Pottery in the Archaeological Record*. Cambridge, Cambridge University Press.

Wylie, A. (2017) How archaeological evidence bites back: strategies for putting old data to work in new ways. *Science, Technology, and Human Values* 42.2, 203–225.

Chapter 11

Evolutionary design processes in thermal architecture of the Roman Empire

Allyson McDavid

Introduction

Longstanding, monumental architecture owes much of its resilience to serial evaluation over its lifetime. Naturally, such assessment would occur at critical stages of building design, maintenance and, at times, adaptation. Historically, however, sustainable practices in architecture have largely been substantiated by episodic evidence of repair: the identification of renewed building fabric simply affirms that consequential decisions were made, at some point in time, to extend the viability of these monuments. But, although regenerative processes of repair, reuse and recycling have similarly defined Roman civic architecture's remarkable longevity, little has been discussed in terms of how, in what sequence, and perhaps most importantly, why these circular practices were engaged beyond the pursuit of structural rehabilitation (Fig. 11.1).

Additionally, when architectural resilience is considered over the longer term and across consecutive phases of design, conservation and even transformation, some traditionally unheralded stakeholders begin to appear alongside the readily visible evidence of sustainability. Who were these participants at various junctures in the construction process, and did their roles change over time? Why were they necessary? Could architectural commissions proceed without them? These otherwise invisible agents may be found at intersections of patronage, design and ceremony of architecture; their mutual dependency upon successful, sustainable outcomes renders their association along the spectrum of architectural practice, crucial – from conception through execution, in ongoing maintenance and in decisions regarding adaptation. By making the social, political and economic mechanisms that underpin sustainable design practice more transparent, trajectories of cultural resilience emerge that are equally as significant to defining architectural circularity as forensic material analysis has heretofore been.

Fig. 11.1 Baths of Antoninus, Carthage, Tunisia. 2nd century AD (Photo by Allyson McDavid).

While the rich, extant remains of *thermae* (imperial-era thermal complexes) allude to their remarkable physical resilience during antiquity, a series of habitually obscure design processes were also in motion over their lifetimes that were not only responsible for ensuring long-term structural durability but also that of the broader, civic institution of Roman bathing. This paper focuses on those less visible collaborations within architectural practice that made these baths' resilience possible: it will variously address ecological incentives for renovation, building patronage and urban administration, design and construction innovation, and ritual evolution. It identifies a commitment to conserve these large baths through Late Antiquity and suggests there was a deliberate ethos of circularity in operation in the ancient world – to an extent that may be surprising to postmodernity.

Challenges to imperial urban beauty in the 3rd century

Throughout classical antiquity, gifted orators associated virtue with the design, construction and maintenance of wholesome urban landscapes (Fig. 11.2). They described the visible to convey the significance of an honoured city: apart from the poetics of a fertile, natural topography that encouraged its foundation, a city's *kallos* (beauty) was particularly expressed in its constructed monuments (Saradi 1995, 37–56; Mass 2005, 89–90; Thomas 2007, 107–126; Webb 2009, 156).[1] From as early as the 1st century AD, civic architecture was considered key to the definition of urban sophistication; without the amenities of temples, theatres, basilicas, stadia and public baths, a city could not be considered beautiful (Suetonius (1st century AD), 29; Pausanias (2nd century AD), 10.4.1; Procopius (6th century AD), 1.10, 4.10.20–23, 5.2.3–5, 4.15–17, 6.4.1–5). Furthermore, these authors equated the physicality of the city to distinctly intangible concerns – namely, the health and wellbeing of citizens who inhabited the urban environment (*e.g.*, Libanius (4th century AD), see Downey 1959, 652–686; and Evagrius Scholasticus (6th century AD), see Bidez and Parmentier 1898, 8). Thermal architecture certainly played a preeminent role in ensuring classical urban magnificence; but compared to other civic monuments, its mandate to engender salubrious living became an urban

Fig. 11.2 'The Course of Empire: The Consummation of Empire,' Thomas Cole, 1836 (Public domain).

prerogative of the imperial period (Fagan 1999, 75–84; Yegül 2014, 299–323). These grand *thermae* singularly embodied *Romanitas*; and their palatial settings became a leitmotif for civic entitlement that assured superiority over less fortunate, less beautiful cities that were absent this luxury (Newlands 2002, 199–226; Fig. 11.3).[2] Such an attitude was buoyed by the decorative and performative spectacle of these thermal environments, which brought distinction to cities that could afford to underwrite the high cost of their operation.

By the late 3rd century, however, Roman urbanites' heretofore privileged, seemingly unassailable quality of life had grown uncertain: a result of more than half a century of relentless civic upheaval, tribal incursions across the Empire's borders, environmental challenges due to uncontrolled resource exploitation, natural disasters, drought, and the corollary onslaught of famine and disease (Fiorino 2009, 261–275).[3] These serial misfortunes did not go unnoticed. During this challenging time, Cyprian, Bishop of Carthage, expounded upon the dire ecological state of affairs in his communication to the proconsul of Africa (Hekster 2008, 130–133):

> The world itself is now announcing, and bearing witness to its decline by the testimony of its failing estate. In the winter there is not such an abundance of showers for nourishing the seeds; in the summer the sun has not so much heat for cherishing the harvest...The layers of marble are dug out in less quantity from the disemboweled and wearied mountains; the diminished quantities of gold and silver suggest the early exhaustion of the metals, and the

Fig. 11.3 'The Baths of Caracalla (3rd century AD),' Lawrence Alma-Tadema, 1874 (Public domain).

> impoverished veins are straitened and decreased day by day; the husbandman is failing in the fields...[as are] skilfulness in the arts, [and] discipline in morals.
>
> Thus...the fountain which once gushed forth liberally from its overflowing veins, as old age causes it to fail, scarcely trickles with a sparing moisture. This is the sentence passed on the world...that everything that has had a beginning should perish...and that strong things should become weak...and that, when they have become weakened and diminished, they should come to an end (Cyprian To Demetrianus, AD 252–257).

While both correspondents were acutely aware of systemic failures in agriculture, trade, mining, infrastructure and artistic creativity across the Roman provinces, they relied upon their religious polarity to explain the cause of these calamities, each attributing precipitous decline in their way of life to the other's heretical affiliation – the bishop to his Christian faith, the proconsul to polytheism. While rhetoric of this kind was prone to hyperbole, grim consequences were already evident, whether or not their fundamental causes were accurately known at that time. Certainly, Roman civilisation as they had known it was no longer sustainable (Temin 2006, 133–151; Erdkamp *et al.* 2015; Van Limbergen and Taelman 2021). To avert catastrophe, therefore, the environment, ecology and infrastructure would need to be managed differently: these now vulnerable, interrelated systems required alteration – either through reinvigoration or more extensive reinvention – to successfully overcome contemporary challenges of the 3rd century. Consequential decisions therefore had to be made, not only to mitigate the effects of unremitting resource exploitation that had been the *modus operandi* of the recent past but also to reverse a trajectory of deprivation that would not soon be forgotten.

The variable fate of Imperial-period *thermae*

It was within this interconnected social, political and environmental context that thermal infrastructure proved a particular challenge to maintain. Of the class of lavish Imperial-period *thermae* in particular, their sheer enormity and dependence on vast quantities of water and fuel caused significant financial strain on cities during the arduous 3rd century. Understandably, many of these baths contracted to a fraction of their foundational size, due to scarcity of natural resources and prolonged economic hardship borne by the cities in which they were situated. Trade inconsistency and episodes of mass deforestation across the empire led to both unreliable fuel supplies and exorbitant heating costs for baths; increasingly frequent droughts combined with ageing, failing aqueducts made water provision a challenge when it had once been considered a limitless, readily accessible resource (Van Sickle 1930, 10; Harper 2017, 129–136; Graham and Van Dam 2019, 330–341. See also the rebuttal to Harper 2017 on citing ecological causes for societal collapse in Haldon *et al.* 2018, 1–13).

Under these conditions and by the 4th century, numerous *thermae* were ultimately abandoned in their entirety; those valiant few cities whose imperial baths were still, even nominally, operational by this time, often elected to convert them into alternative facilities as they could no longer continue to function as they had been designed to do. These baths were variously transformed into Christian temples, prisons or factories for pottery and glass production; their impressive structures were conducive to a range of burgeoning Late Antique needs from worship to incarceration, while furnaces and water conduits that were purpose-built for bathing could instead transfer their value to industrial ends (Bouras 1981, 643–644; 2002, 525–526; Berger 1982, 1–172; Mango 1982, 338; Karpozilos *et al.* 1991, 271–272; Russell 2002, 221–228; Lavan *et al.* 2007, 156, 307). In far more numerous cases, however, *thermae* were left entirely to decay in the 4th century by cities no longer able, or willing, to either restore them as baths or commit them to alternative, useful purposes. The materials of their construction were instead quarried and repurposed for cities' more urgent needs – including fortification, repair of priority infrastructure, housing, etc. (Kinney 1986, 379–397; 2006, 233–252; Meneghini and Santangeli Valenzani 2004, 179; Coates-Stephens 2016, 184–199; Dumser 2018, 147–148).

During the High Imperial period, and no later than the end of the Severan dynasty in AD 235, 53 great *thermae* were founded across the Empire (Nielsen 1993, 1, 98; Thébert 2003, 2, 411; Saliou 2004, 295). Curiously, however, a significant number from this imperially-founded group – some 17 large baths – were converted into 'summer baths': seasonally operable thermal complexes that continued to function into Late Antiquity, seemingly in defiance of the trajectory of decline for so many *thermae* that were either left to ruin or were converted to other uses by the 4th century (Thomas and Witschel 1992, 135–177; Saliou 2004, 301–306). Of these, only five extant structures are securely identified: at Thuburbo Majus and Sbeïtla in Tunisia; at Madaurus in Algeria; in the city of Fordongianus, Sardinia; and at Aphrodisias in Turkey (Fig. 11.4; McDavid 2019, 1–19). These seasonal baths were uncommon for providing bathing facilities, without closure, from their inception during the High Imperial Period through Late Antiquity, when insurmountable ecological factors precipitated their final abandonment and that of entire urban centres more extensively; all-too-frequent narratives relate the inability of cities to rebuild after catastrophic natural disasters from the 7th century on. Although some 23 large baths were newly built across the Roman Empire between the 4th and 6th centuries,

Fig. 11.4 Caldarium, Summer Baths of Sbeïtla, Tunisia. 4th century AD (Photo by Allyson McDavid).

none of the summer baths of this period were new foundations. Instead, significant investment was made by particular Late Antique city administrations, donors and building professionals to rehabilitate those preexisting (albeit aged), Imperial-era thermal foundations. Accompanying these decisions to renovate was an awareness that the culture of public bathing had to change, significantly, to endure in Late Antiquity – a consequential period of transition for Roman society.

The material evidence of resilience for this specific group of summer baths – some that functioned, astonishingly, for up to half a millennium – certainly provides insight into how these extraordinary examples of longevity were made possible. But perhaps more importantly, one might ask: why were these buildings tasked to function past the 3rd century, when a culture of unmitigated resource consumption – ostensibly the *raison d'être* of these behemoths – proved unsustainable? What prompted longstanding systems to evolve to preserve these baths' value, particularly given the pressing social, economic and ecological exigencies of Late Antiquity? And from these causes, what other patterns of circularity might be implied for Late Antique thermal infrastructure, beyond the tangible material of repair?

Invisible causality of repair

We do not have knowledge of a thing until we have grasped its why, that is to say, its cause.
– Aristotle, *Physics*, 4th century BC

Repair is something of a paradox: it represents both presence and absence (Ng and Swetnam-Burland 2018). In architecture, as an aim to cauterise effects of decay, failure or redundancy, repair is not only remedial but it also erases evidence of failure; depending on the ability to distinguish the new from pre-existing, unaffected building fabric, the excision of redundant components renders failure both invisible to the eye and absent from public memory. Furthermore, repair is often overlaid by decorative finish, which obscures the corrective work done to systems embedded beneath the structure's visible surface (Fig. 11.5). But certainly, these finishes define the aesthetic culmination of otherwise invisible, seemingly unspectacular work of systems restoration. In addition, repair can readily be signified by wholesale replacement, but it may also involve more subtle refinement within larger systems infrastructure to improve overall building performance. In the example of thermal architecture, these otherwise invisible systems included those devised for heating, water and air distribution – all of which involved the management of natural resources.

Historic repairs underscore the passage of time for monuments; when viewed individually, these episodes of material corrosion or structural debilitation may be seen as distinct, unrelatable events (*histoire événementielles:* Lacombe 1894; Simiand 1903, 1–22). Over the *longue durée*, however, the solicitation of repair – when considered in serial and typological terms – becomes consequential to a narrative of architectural sustainability (Braudel 1980, 64–65). Viewed in their totality, these repairs invite us to ask broader questions of causality that not only encourage the synchronisation of

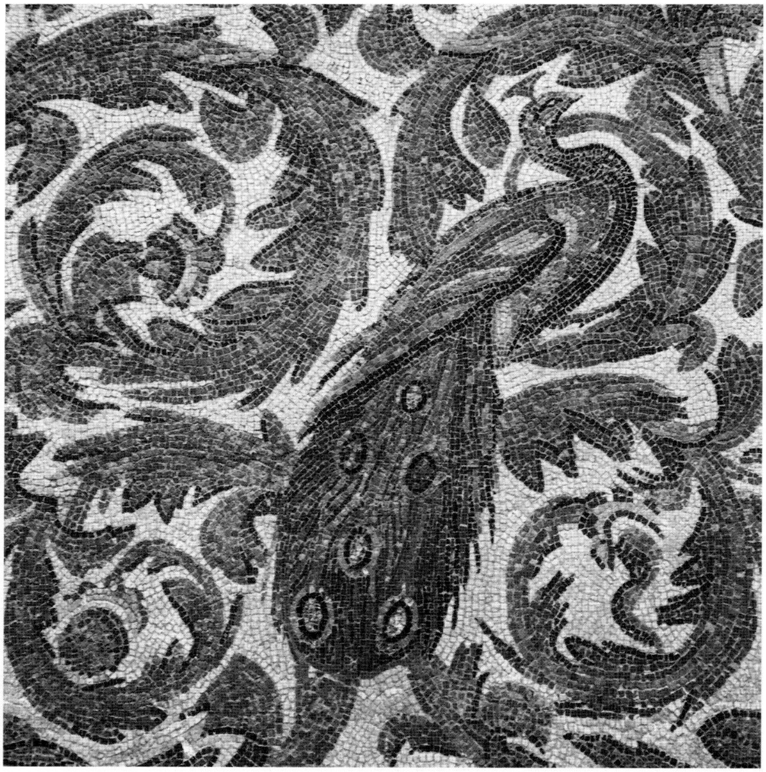

Fig. 11.5 Mosaic decoration. Frigidarium, Summer Baths of Thuburbo Majus, Tunisia. AD 408 (Photo by Allyson McDavid).

discrete occurrences of alteration, but also help to suggest a rationale for why changes were commissioned at all. That relationship between individual cause and its effect, and across correlated phenomena, has been the subject of profound consideration since antiquity (Aristotle, *Physics* II.3).[4] Answers to why repairs were commissioned may be pursued through evidence of matter, design, purpose and agency; these four causes are neither mutually exclusive, nor always visible (Heidegger 1977, 289–290).[5] As societal needs are also prone to change over time, such acts of repair, alteration, renovation or even transformation may be seen as iterative processes of institutional continuity; a protracted view of the culture of repair may therefore reveal otherwise invisible, evolutionary changes in community ideals over time. Of these four fundamental explanations or causes, archaeology has largely concerned itself with three of them: those of material, form and function (efficiency). The fourth aspect, agency – the fundamental generator of change – has been studied very little.

Agents in the Roman building profession encompassed design practitioners and those who sourced, transported and assembled materials; the work of the *architekton* (master builder), masons, bricklayers, carpenters, pipefitters, furnace operators, painters, mosaicists, etc. is clearly visible in the vast corpus of archaeological ruins.

But other agents who were less visible, but as essential as those involved in hands-on construction, were patrons who underwrote commissions for restoration, city administrators who acted as conduits between philanthropists and contractors, those who attended to procurement and distribution of funds, and people who inhabited these venues. For thermal architecture, the four causes of repair represented critical stages in the provision of bathing culture; when observed over an extended period of time, the least visible cause – that of agency – becomes critical to identifying circularity. But the visible archaeological material may be effectively used to address those unseen systems through which thermal architectural form was activated and engaged, and which enabled some baths to continue to function for a remarkably long time. Indeed, these invisible processes were vital to the sustainability and necessary evolution of these baths, from their imperial foundation to Late Antique modification.

Traditionally invisible agents of economic circularity may be identified by the roles they played within the broad spectrum of architectural activity – from conception to completion. These parties include patrons, whose interests extended beyond mere civic duty; urban administrators, whose mandate included maintenance of civil society; and masons, artisans and craftsmen whose skills were put to executing multiple, often sequential building commissions. Through their distinction, cycles of circularity appear that illustrate change in each of their trajectories over time; they are manifest, respectively, through transition in the source and scope of patronage, evolution in the dynamics of urban administration, and innovation in masonic skill. Masonic skill is visibly demonstrated in the archaeological remains and has been previously discussed at length (Ousterhout *et al.* 2012; McDavid 2019); this paper instead addresses less visible processes within architectural patronage and administration, systems regulation and building use.

(i) Urban administration: patronage

While on-site inscriptions primarily commemorated major commissions for building repair, all baths underwent regular servicing as often as every 10 or 20 years to counter ill-effects of extensive foot traffic, standing and running water, steam and extremes of temperature on the building fabric (Rebuffat 1970, 107–121). These repairs included light building maintenance, as well as cosmetic or decorative repair in the form of repainting, plasterwork and refinishing of floors and walls – all matters of course for baths. But while such quotidian campaigns of renovation were not publicly celebrated (Thomas and Witschel 1992, 159), they were documented in surviving papyrus invoices from Roman Oxyrhynchus in Egypt, written during the 4th to 6th centuries between city administrators and contractors from a myriad of building trades. These invoices were never seen by the Late Antique public, but they paint a portrait of meticulous decision-making to retain baths in working order (Grenfell and Hunt 1898): they recorded costs for painting, lime plaster preparation and its delivery (P.*Oxy* XII 1104; AD 306), delivery of charcoal for heating (*P.Oxy* XII 1430; AD 324), payment for lead and tin to repair pipes (P.*Oxy* VI 915; AD 572) and the budget for replacing window glass

Fig. 11.6 Inscription ALA78, *'Third Honors for Rhodopaios,' Hadrianic Baths of Aphrodisias, Turkey. AD 542 (From* Aphrodisias in Late Antiquity *online database).*

(*P.Oxy* XII 1430; AD 584). Such periodic, small-scale renovation campaigns were *de rigeur* and paid for from cities' annual maintenance budgets. These repairs often erased the visibility of corrosion, breakage or redundancy, and a visitor to the baths would be solely aware of decorative finishes and performance, rather than of internal, structural interventions.

Larger-scale, transformative renovations, however, were publicly commemorated by epigraphic dedications on par with those celebrating a building's earlier foundation (Fig. 11.6). During the High Imperial period, urban elites competed to finance repairs of cities' most noteworthy monuments in their names, often with an aim to eclipse their social peers (Robert 1978, 64–67, 74–77; Roueché 1984, 181–188; Whittow 1990, 6). For families of the *curial* (senatorial) class in western Asia Minor, networks of building patronage even extended beyond a single city: the Licinii, for example, supported renovation throughout Lycia, Phrygia and Caria, across which numerous commemorative statues were also erected in their honour (Kearsley 1988, 43–51). Late Antique inscriptions, similar in earlier commemorative style, additionally alluded to structural alteration, reconfiguration of water features and conduits, changes to pedestrian circulation, wall modifications and repair of floor and wall-heating systems; while many were inscribed on architectural surfaces in proximity to work executed, some annotated the freestanding, sculptural portraits of donors who facilitated these transformative commissions. None of the large-scale renovations would have been possible without such sponsorship, and Late Antique benefactors looked to tradition to continue the practice of public sponsorship and personal aggrandisement.

But by the 3rd century, economic fluctuation and structural crises had undermined 'old attitudes of thought and action, resistant frameworks dying hard, at times against all logic,' in civic affairs (Braudel 1980, 25–54). Across the Empire over the ensuing 4th to 6th centuries, the seemingly hermetic political status of provincial, longstanding curial families began to deteriorate. Having previously commanded power of patronage among their select ranks, by Late Antiquity, new and quite different epicentres of largesse emerged to fund building projects: from landowners, a burgeoning Christian clergy and career professionals such as doctors, philosophers and local administrators (Roueché 1984; 2004).[6] The political environment within which imperial *thermae* were a consistent focus of patronage for over half a millennium had changed dramatically during that time. Epigraphic evidence for this evolving trajectory suggests that while there was continued agency in patronage to maintain urban infrastructure, capital was being raised from outside traditional spheres of aristocracy during Late Antiquity.

(ii) Systems circularity and resource management

Ecological circularity may be traced through the management of bathing infrastructure that harnessed water and fuel resources. Architecturally, these goals were met in the design of heating systems embedded within wall cavities that included furnace conduits and terracotta *tubuli* (box flues); in underfloor, heated gas distribution systems framed by *suspensurae* (suspended floors) and hypocausts; and in the provision of plumbing that fed fountains and *alvei* (immersion pools) (Fig. 11.7). These heating, convection and water systems were customarily hidden beneath decorative veneer to render them invisible to the bather's eye; but the substantial evidence of their alteration (gained through exposure of their contemporary ruins) strongly suggests both a late Roman concern for continuity in the delivery of essential bathing amenities and an awareness of the value of natural resources during Late Antiquity, as their procurement and distribution became increasingly precarious.

Heating required the greatest investment for baths, both in terms of the consumption of natural resources and in the development of infrastructure to manage their deployment. Epigraphic surveys across the Empire during the High Imperial period attest to projected heating costs as a major consideration for cities, being specifically earmarked in their annual budgets for the maintenance and upkeep of baths. For large *thermae*, these were staggering numbers of up to 400,000 *sestertii* annually for its wood fuel; another site, though absent financial detail, required 400 cartloads of hardwood per year and indicated that this was the main expense for these baths (Duncan-Jones 1965, 204, 208). Such an expense became unsustainable for most during the 3rd century; for these baths to remain viable, the demand for and consumption of fuel had to

Fig. 11.7 Hypocaust system, Hadrianic Baths of Aphrodisias, Turkey, 2nd–7th century AD (Photo by Allyson McDavid).

be greatly reduced, necessitating the overhaul of their heating systems. In addition, alternative fuel sources were found in Late Antiquity through burning leftover waste from olive oil manufacture in thermal furnaces instead of wood; it was an inspired method of circular reuse, rather than simple disposal of the by-product or its incineration elsewhere (Rowan 2015, 465–482).

Tubuli – the earthenware, box-shaped tiles used to channel heated air through the walls of baths – distinguished heated rooms from non-heated ones and were standard components of bathing infrastructure since Seneca's lifetime in the 1st century AD (*Sen. Ep.* 86; Yegül 1992, 40). By the High Imperial period, these terracotta flues were produced in large quantities and exported from Rome and other major distribution centres across the Empire to aid in the rapid, planned construction of baths during this era (Maréchal *et al.* 2021, 243–247). Their standardised dimensions assumed a high volume of airflow through their cubic voids, with equally copious fuel consumption to produce the heated air intended to pass through them. For these *tubuli*, apart from their high intake of heated air at any given time, there was also an unquestioned expectation that wood, critical to heating that air, was readily accessible.

Fig. 11.8 Wall convection system, Hadrianic Baths of Aphrodisias, Turkey, 2nd–7th century AD (Photo by Allyson McDavid).

For baths that endured past the 4th century, however, the profligate use of wood and capacious air circulation became indefensible. Alternative and more cost-effective methods for convection needed to be developed to constrain the use of natural resources that had been unrelenting during the High Imperial period. In response, terracotta components began to be produced locally in Late Antiquity, eliminating the need to import *tubuli* that had adversely impacted city finances through their long-distance trade and use (Tomber 1987, 161–174). From the 4th century on, at least four distinct terracotta flue systems were developed to facilitate vertical convection in Roman baths (Yegül 1992, 364; Adam 2005, 557–561; Koçyiğit 2006).[7] These types were often employed serially within a single complex, suggesting an accretion of several campaigns of repair over time (Fig. 11.8). Where these overlays were visible, the void once created by foundational *tubuli* was occasionally filled in with mortar, rendering it non-convectional but still able to retain and conduct heat (Yegül *et al.* 1986, 8, 115; Horsley 1987, 49–80; McDavid 2016, 222–223). The regional manufacture of components fostered a bespoke, iterative approach to heating systems design: as critical decisions in the construction of wall cavities could now be made at the local level, the subsequent operational refinement in baths equipped those cities to reconcile fuel expenditure more responsibly. Though perhaps chiefly motivated by fiscal prudence, Late Antique design innovations in heating and convection helped to advance cities' efforts in resource conservation, especially when compared to their outlay using High Imperial heating systems.

Heated *alvei* (immersion pools) were also prone to the vagaries of resource management, as they relied upon wood-burning furnaces abutting the rooms in which these pools were situated. In these publicly accessible *tepidaria* and *caldaria*, some of which were heated to temperatures of up to 57°C (Kretzschmer 1961, 91), their heated pools were a cornerstone of bathing activity: while meeting the exigencies of becoming clean, they additionally provided tactile, auditory and visual enjoyment. During the High Imperial period, these pools were typically designed

Fig. 11.9 Blocked furnace arch, Hadrianic Baths of Aphrodisias, Turkey, c. 4th century AD (Photo by Allyson McDavid).

to be large, deep and luxurious in their provision of hot water. By Late Antiquity, however, several of the furnace arches that directly linked these immersion pools were entirely blocked or had their apertures drastically reduced (Yegül 1992, 257, 281, 370; McDavid 2019, 8; Fig. 11.9).[8] Though embedded in the wall envelope and invisible to the eye, their deliberate alteration indicates that delivery of heat was no longer permitted. Many of the summer baths' furnace arches exhibited this particular alteration, with such repair likely precipitated by economic pressures of high fuel costs. Although the pools once fed by these altered furnace arches did continue in their immersive function, several of them effectively became non-circulating pools; this loss of convection capability resulted in pools having to raise their ambient temperature manually or else forego attempts at heating altogether. But visitors would not have missed heated pools during the temperate months these baths were operational – certainly not to the high temperatures to which they were originally heated – and the enjoyment of some measure of traditional, shared immersion that had been the *status quo* from the 2nd century on, was maintained into Late Antiquity.

Accompanying the demand to reduce heating costs, several small and entirely unheated pools were newly installed in bathing rooms after the 4th century (Ginouvès 1955, 141–416; Biers 1985, pl. 38; Yegül 1992, 329–339; Charpentier 1994, 113–142; Fig. 11.10).[9] As part of a widely attested development across the eastern provinces by the 5th century, these small pools were particularly useful in the longstanding summer baths, as increased access to cool immersion was desirable during the warm months of the year. The numerous interior, unheated and non-circulating pools built into the older frames of the baths in Thuburbo Majus, the *frigidarium* at Sbeïtla, at Fordongianus and at Aphrodisias signalled similar concern for more careful resource management and expenditure (Biers 2003, 315). Additionally, small pools were installed in various exterior locations about the summer *thermae*; these were also designed for unheated use only and were enjoyed throughout the approximately nine months per year these complexes were in use.

Fig. 11.10 Indoor pool, Forum Baths at Fordongianus, Sardinia. c. 5th century AD (Photo by Allyson McDavid).

(iii) Ritual circularity: use and performance

Change in social rituals of undress, bathing and massage at summer baths was evident in the alteration of pedestrian circulation through these spaces, from their imperial foundation into Late Antiquity. One significant example at Aphrodisias is defined by the closure of the *caldarium's* original, 2nd-century, monumental arches that provided access to the building's central heated core. In the building's final, 7th-century form, however, two small pools had been built into these entries that blocked egress; simultaneously, four alternate passages were carved out of the *caldarium's* apsidal niches that flanked the arches, that once housed either monumental marble sculpture or cold-water basins to provide bathers with relief from the intense heat of the room. Not only was access to the Baths' ultimate destination changed but also the ceremony of arrival that had been established in the Imperial Period. The construction of multiple entries into the *caldarium* would have challenged the room's ability to retain heat during Late Antiquity, as their inclusion would have increased the room's rate of transpiration: an adverse outcome if the high temperature of the caldarium was to be retained. But, given cities' burdens with heating costs, it was likely that *caldaria* were no longer tasked to attain the high temperatures they achieved in the 2nd century and instead maintained a more economically manageable level of ambient warmth (Ginouvès 1955, 147–152).

The *caldarium* at the baths at Thuburbo Majus, Tunisia, was divided into several parts that allowed multiple opportunities of egress in Late Antiquity, as was similarly the case at Aphrodisias. This division enabled these rooms to be used on the one hand by smaller constituencies, whereby a single sector of the large room could be heated at a given time and the remainder of the room left unheated; on the other hand, different kinds of people could have been served at one time, while remaining distinguished within the space. While gender separation has historically been defined as a pressing societal issue of Late Antiquity in textual sources, the newly Christianised empire would also have demanded very different rules of engagement between men and women at baths; this spatial division found in summer baths gives a strong rationale for gendered ceremonial within thermal precincts that was undetectable apart from inscriptional evidence (Dunbabin 1989, 6–7; Ward 1992, 125–26; Arapoyianne 1995, 155).

These divided spaces also lent themselves to controlled management of heating and water resources while they facilitated gender separation. There also may have been a concern to accommodate emerging preferences for individualised bathing. The small pools added to the complexes offered a somewhat exclusive, private amenity within the larger public venue, that was a significant departure from the communal bathing activity originally intended in the Imperial period. These spatial alterations were significant contributors to both architectural and cultural resilience as economic, gendered and religious imperatives became paramount for Late Antique society in transition from its High Imperial heyday (Ginouvès 1955, 133–147).

The summer baths evolved along a sacramental trajectory as well as a temporal one. Given the substantial survival of pagan statuary within bath precincts into

Fig. 11.11 Ex-voto *amulet dedicated to Asklepios from the caldarium* (ALA *63.568), Hadrianic Baths of Aphrodisias, Turkey. c. 2nd century AD (From* Aphrodisias in Late Antiquity *online database).*

Christianising Late Antiquity, such ritual inclusivity suggests gradual transition and an openness to preservation of its past (Bloch 1977, 282; Fogelin 2007, 58). Apart from the maintenance of monumental and life-sized statuary of both mythical and imperial themes within these complexes for decorative purposes, many portable, *ex-voto* artefacts dedicated to Asklepios continued to be used in bathing spaces during Late Antiquity; the presence of less portable stone altars, identified within several rooms, further distinguished these complexes for their religious ceremonial in addition to the other, more quotidian ritual of bathing (Fig. 11.11). Numerous Christian symbols that were sculpted, etched or painted, were also found in the vicinity of immersion pools, implying a plurality of worship at these venues during Late Antiquity. By AD 513, references to Christian Monophysite controversy were made in inscribed, painted fresco decoration on the walls of the north *tepidarium* of the Hadrianic Baths at Aphrodisias; from these, an emergent Christian faith appears to have wielded growing political influence through sponsorship of the baths' restoration, one that gained a highly visible platform within a setting typically ascribed to the secular. This varied religious engagement illustrates transition in ritual practice over a prolonged period of time from pagan to ecumenical, that embraced Christian ascendancy within a remarkably pluralistic ritual landscape.

Conclusion

The undertaking of architectural practice that made design, construction and repair a success constituted several agents: civic sponsorship, urban administration, resource management, masonic skill and systems innovation. Their collective participation not only secured the structural integrity of summer baths for some half a millennium but they also served to uphold longstanding thermal rituals at a time when this tradition became increasingly prohibitive to maintain. In recognising that long-held institutions often required reinvention to endure, this paper viewed the archaeological material of repair as a final, visible outcome of many stages: the end of a process that was only successful through activation of various social, economic and ecological practices – ones that are often forgotten once acts of repair are complete. But threatened with the almost unthinkable loss of *Romanitas* itself in the demise of the Empire's imperial *thermae*, diverse urban patrons, design professionals, masons and resident citizenry took drastic steps to retain, in some measure, the culture of public bathing in the face of a fundamentally changing society. Where those 'invisible' mechanisms were leveraged, they successfully demonstrated the priority baths held in their cities and facilitated thermal architecture's long-term sustainability through Late Antiquity. It additionally appears to be the case that the longstanding institutions of patronage, administration and the building arts themselves required reinvention over time to endure; in defining their evolution using the archaeological evidence of baths, these otherwise invisible practices of architectural circularity may broadly apply to cities across the Roman world, in their uncompromising pursuit of immortality.

Notes

1 For a timeline of rhetorical *encomium urbis/urban descriptiones* (urban eulogies), see Menander Rhetor and Dionysius of Halicarnassus, *Ars Rhetorica*, AD 285–312; Libanius, *Antiochicus* (*Oration in Praise of Antioch, Oration XI*), *c.* AD 360; Ausonius, *Ordo Nobilium Urbium*, 4th century AD; Rutilius Namatianus, *De reditu suo*, early 5th century AD.

2 Notable ekphraseis on baths: Statius, *Silvae* Book I:5 'The Baths of Claudius Etruscus' (1st century AD); Martial 6.42 and 9.75 (1st–2nd century AD); Lucian, *Phalaris Hippias or The Bath* (2nd century AD); Favorinus, *On Bath-houses* (2nd century AD); Philostratus *Vitae Sophistarum* 1.8 (3rd century AD); Menander Rhetor and Dionysius of Halicarnassus, *Ars Rhetorica* 365.18–22 (3rd–4th century AD). For an in-depth overview of Roman bath expository, see Busch 1999.

3 The Plague of Cyprian (AD 249–262), descending on the heels of the Antonine Plague/Plague of Galen (AD 165–180), led to severe population decline that included the military ranks and severely undermined the Roman Empire's ability to repel tribal migration and incursion; it also led to widespread famine, as many farms were abandoned or became unproductive under these trying conditions. Barbarian migration and invasion were also likely due to consistent summers of drought (with famine a consequence), a phenomenon we today consider a result of climate change.

4 In Aristotle's use of the term αἰτία – the 'explanation' or 'answer to a "why" question' – the four causes are either material, formal, efficient or final.

5 Heidegger echoed Aristotle in a similar exploration of causality, with his *causa efficiens* (the effect that brings about the finished result), a mirror of Aristotle's 'agency,' being crucial to understanding why changes occur.

6 For Aphrodisias: Helladios, a governor (*ALA* 17, 18); Dionysios, a doctor (*ALA* 67), Rhodopaios, a grain distributor (*ALA* 85, 86, 87).

7 Convection components employed in Asia Minor from the 4th century AD: (i) tubuli; (ii) *tegulae mammatae*; (iii) *Ton-nägeln* (terracotta studs); and (iv) iron studs with terracotta spacers.

8 *E.g.*, the Kaiserthermen at Trier, the theatre bath-gymnasium at Ephesos, the Hume-i Tepe baths at Miletus and the Hadrianic Baths at Aphrodisias.

9 Syria: Androna (*c.* AD 558), Sergilla (AD 473), Babisqa (AD 488); Greece: East Baths at Delphi, Thermes A at Argos, Northeast Baths at Epidauros; Asia Minor: Harbor Baths at Ephesus, Imperial Baths at Caunos, Imperial Baths at Sardis.

Bibliography

Primary sources

Aristotle. *Physics*. Translated by Hardie, R.P. and Gaye, R.K. *Physics*. In Jonathan Barnes (ed.) *Complete Works of Aristotle*, vol. 1, 315–446. Princeton, Princeton University Press, 1984.

Aristotle. *Posterior Analytics; Topica*. Edited by Hugh Tredennick, Edward S. Forster. Cambridge, Harvard University Press, 1960.

Evagrius Scholasticus, *Ecclesiastical History*. Edited by Michael Whitby. *The Ecclesiastical History of Evagrius Scholasticus*. Liverpool, Liverpool University Press, 2000.

Favorinus, *On Bath-houses*. In Dio Chrysostom. *Discourses, 37-60*. Edited by H. Lamar Crosby. Cambridge, Harvard University Press, 1971.

Lucian, *Phalaris Hippias or The Bath*. Edited by Austin M. Harmon, *Lucian* vol.1. Cambridge, Harvard University Press, 1913.

Martial, *Epigrams*. Edited by Walter C.A. Ker. London, Heinemann, 1919.

Menander Rhetor and Dionysius of Halicarnassus. Edited by William H. Race. *Menander Rhetor. Dionysius of Halicarnassus, Ars Rhetorica*. Cambridge, Harvard University Press, 2019.

Pausanias, *Description of Greece*. Edited by William H.S. Jones. *Description of Greece Volume I: Books 1-2 (Attica and Corinth)*. Cambridge, Harvard University Press, 1918.

Philostratus, *Vitae Sophistarum*. Edited by Rudolf Stefec. *Lives of the Sophists*. Oxford, Oxford University Press, 2016.

Procopius. *On Buildings*. Edited by Henry B. Dewing. Cambridge, Harvard University Press, 1940.

Suetonius. *Divus Julius - The Lives of the Twelve Caesars*. Translated by Alexander Thomson, J. Eugene Reed. Philadelphia, Gebbie & Co, 1889.

Secondary sources

Adam, J.-P. (2005) *Roman Building: Materials and Techniques*. London, Routledge.

Arapoyianne, X. (1995) Τὸ ρωμαΐχὸ βαλανεῖο στὴν Μπρεξιζα τοῦ Μαραθῶνος. *Archaiologike Ephemeris* 132, 133–186.

Berger, A. (1982) *Das Bad in der byzantinischer Zeit*. Munich, Institut für Byzantinistik und neugriechische Philologie der Universität.

Bidez, J. and Parmentier, L. (1898) (eds) *The Ecclesiastical History of Evagrius*. London, Methuen.

Biers, J. (1985) *Corinth vol. XVII: The Great Bath on the Lechaion Road*. Princeton, Town House Press.

Biers, J. (2003) Lavari est vivere: baths in Roman Corinth. *Corinth* 20, 303–319.

Bloch, M. (1977) The past and the present in the present. *Man*, 12.2, 278–292.

Bouras, C. (1981) City and village: urban design and architecture. *Jahrbuch der österreichischen Byzantinistik* 30.1, 611–653.

Bouras, C. (2002) Aspects of the Byzantine City. In A. Laiou (ed.) *The Economic History of Byzantium*, 525–526. Washington D.C., Dumbarton Oaks.

Braudel, F. (1980) *On History and the Social Sciences: The* Longue Durée. Chicago, University of Chicago Press.

Busch, S. (1999) *Versus balnearum: Die antike Dichtung über Bäder und Baden im römischen Reich*. Stuttgart, B.G. Teubner.

Charpentier, G. (1994) Les bains de Sergilla. *Syria* 71, 113–142.

Coates-Stephens, R. (2016) *Building in Early Medieval Rome, 500-1000 AD*. Unpublished thesis, University College London.

Downey, G. (1959) Libanius' oration in praise of Antioch (Oration XI). *Proceedings of the American Philosophical Society* 103.5, 652–686.

Dumser, E. (2018) Reuse as a late Roman building practice. In D. Ng and M. Swetnam-Burland (eds) *Reuse and Renovation in Roman Material Culture: Functions, Aesthetics, Interpretations*, 147–148. Cambridge, Cambridge University Press.

Dunbabin, K.M.D. (1989) Baiarum Grata Voluptas: Pleasures and dangers of the baths. *Papers of the British School at Rome* 57, 6–46.

Duncan-Jones, R. (1965) An epigraphic survey of costs in Roman Italy. *Papers of the British School at Rome* 33, 189–306.

Erdkamp, P., Verboven, K. and Zuiderhoek, A. (2015) (eds) *Ownership and Exploitation of Land and Natural Resources in the Roman World.* Oxford, Oxford University Press.
Fagan, G. (1999) *Bathing in public in the Roman world.* Ann Arbor, University of Michigan Press.
Fiorino, S. (2009) The Antonine Plague and the decline of the Roman Empire. *Le Infezioni in Medicina: Rivista Periodica di Eziologia, Epidemiologia, Diagnostica, Clinica e Terapia delle Patologie Infettive* 17.4, 261–275.
Fogelin, L. (2007) The archaeology of religious ritual. *Annual Review of Anthropology* 36, 55–71.
Ginouvès, R. (1955) Sur un aspect de l'évolution des bains en Grèce vers le IV siècle de notre ère. *Bulletin de Correspondance Hellénique* 79, 133–152.
Graham, B. and Van Dam, R. (2019) Modelling the supply of wood fuel in ancient Rome. In A. Izdebski and M. Mulryan (eds) *Environment and Society in the Long Late Antiquity,* 330–341. Leiden, Brill.
Grenfell, B.P. and Hunt, A.S. (eds) (1898) *The Oxyrhynchus papyri* vol. 6. London, Egypt Exploration Fund.
Haldon, J., Elton, H., Huebner, S., Izdebski, A., Mordechai, L. and Newfield, T. (2018) Plagues, climate change, and the end of an empire: A response to Kyle Harper's *The Fate of Rome (1)*: Climate. *History Compass* 16.2, 1–13.
Harper, K. (2017) *The Fate of Rome.* Princeton, Princeton University Press.
Heidegger, M. (1977) *The Question Concerning Technology and Other Essays.* New York, Harper Collins Publishers.
Hekster, O. (2008) *Rome and its Empire, AD 193-284.* Edinburgh, Edinburgh University Press.
Horsley, G.H.R. (1987) The inscriptions of the so-called library at Cremna. *Anatolian Studies* 37, 49–80.
Karpozilos, A., Johnson, M.J., Kazhdan, A. and Browning R. (1991) Baths. In Alexander P. Kazhdan (ed.) *The Oxford Dictionary of Byzantium,* vol. 1, 271–272. Oxford, Oxford University Press.
Kearsley, R.A. (1988) A leading family of Cibyra and some Asiarchs of the first century. *Anatolian Studies* 38, 43–51.
Kinney, D. (1986) Spolia from the Baths of Caracalla in Sta. Maria in Trastevere. *Art Bulletin* 68, 379–397.
Kinney, D. (2006) The concept of spolia. In C. Rudolph (ed.) *Companion to Medieval Art: Romanesque and Gothic in Northern Europe,* 233–252. Chichester, Wiley.
Koçyiğit, O. (2006) Terracotta spacers from the bathhouse at Amorium. *Anatolian Studies* 56, 113–125.
Kretzschmer, F. (1961) *Die Entwicklungsgeschichte des antiken Bades und das Bad auf dem Magdalensberg.* Düsseldorf, VDI Verl.
Lacombe, P. (1894) *de l'Histoire considerée comme science.* Paris, Hachette.
Lavan, L., Zanini, E. and Sarantis, A. (eds) (2007) *Technology in Transition: A.D. 300-650.* Leiden, Brill.
Mango, C. (1982) Daily life in Byzantium. *Akten XVI. Internationalen Byzantinistenkongress, Akten I/1 = Jahrbuch der Osterreichischen Byzantinistik* 32.1, 337–353.
Maréchal, S., de Haan, N. and Clerbaut, T. (2021) Manufacturing tubuli: an experimental reconstruction of the *chaîne opératoire* based on archaeological evidence. In M. Gnade and M. Revello Lami (eds) *Tracing Technology: Forty Years of Archaeological Research at Satricum,* vol. 42, 243–247. Leuven, Peeters.
Mass, M. (ed.) (2005) *The Cambridge Companion to the Age of Justinian.* Cambridge, Cambridge University Press.
McDavid, A. (2016) The Hadrianic baths in Late Antiquity. In R.R.R. Smith (ed.) *Aphrodisias Papers 5: New Research on the City and its Monuments,* 209–224. Portsmouth, Journal of Roman Archaeology Supplementary Series.
McDavid, A. (2019) Thermal sustainability: renovation of imperial thermae in Late Antiquity. *Architectural Histories* 7.1, 1–19.
Meneghini, R. and Santangeli Valenzani, R. (2004) *Roma nell'altomedioevo. Topografi a e urbanistica della città dal V al X secolo.* Roma, Libreria dello Stato.

Newlands, C. (2002) Statius' *Silvae and the Poetics of Empire.* Cambridge, Cambridge University Press.
Ng, D. and Swetnam-Burland, M. (2018) *Reuse and Renovation in Roman Material Culture: Functions, Aesthetics, Interpretations.* Cambridge, Cambridge University Press.
Nielsen, I. (1993) *Thermae et Balnea. The Architecture and Cultural History of Roman Public Baths,* 2 vols. Aarhus, Aarhus University Press.
Ousterhout, R., Holod, R. and Haselberger, L. (eds) (2012) *Masons at Work: Architecture and Construction in the Pre-Modern World.* Philadelphia, University of Pennsylvania.
Rebuffat, R. (1970) Bu Njem 1970. *Libya Antiqua* 6.7, 107–121.
Robert, L. (1978) Documents d'Asie Mineure. *Bulletin de correspondance hellénique,* 64–77.
Roueché, C. (1984) Acclamations in the later Roman empire: new evidence from Aphrodisias. *Journal of Roman Studies* 74, 181–188.
Roueché, C. (2004) *Aphrodisias in Late Antiquity*, electronic second edition. London, King's College.
Rowan, E. (2015) Olive oil pressing waste as a fuel source in Antiquity. *American Journal of Archaeology* 119.4, 465–482.
Russell, J. (2002) Anemourion. In A. Laiou (ed.) *The Economic History of Byzantium,* 221–228. Washington, D.C., Dumbarton Oaks.
Saliou, C. (2004) Bains d'été et bains d'hiver: Antioche dans l'empire romain. In B. Cabouret, P.-L. Gatier and C. Saliou (eds) *Antioche de Syrie, Histoire, images et traces de la ville antique,* 289–309. Paris, De Boccard.
Saradi, H. (1995) The *kallos of* the Byzantine city: the development of a rhetorical *topos and* historical reality. *Gesta, The International Center of Medieval Art* 34.1, 37–56.
Simiand, F. (1903) Méthode historique et siences sociales. *Revue de synthese historique,* 1–22.
Temin, P. (2006) The economy of the early Roman empire. *Journal of Economic Perspectives* 20.1, 133–151.
Thébert, Y. (2003) *Thermes romains d'Afrique du Nord et leur context méditerranéen.* Études d'histoire et d'archéologie 2 vols. Rome, École Française de Rome.
Thomas, E. (2007) *Monumentality and The Roman Empire: Architecture in the Antonine Age.* Oxford, Oxford University Press.
Thomas, E. and Witschel, C. (1992) Constructing reconstruction: claim and reality of Roman rebuilding inscriptions from the Latin West. *Papers of the British School at Rome* 60, 135–177.
Tomber, R. (1987) Evidence for long-distance commerce: imported bricks and tiles at Carthage. *Rei Cretariae Romanae Fautorum Acta* 25/26, 161–174.
Van Limbergen, D. and Taelman, D. (2021) The exploitation of raw materials in the Roman world: A closer look at producer-resource dynamics. In D. Van Limbergen and D. Taelman (eds) *Archaeology and Economy in the Ancient World - Proceedings of the 19th International Congress of Classical Archaeology, Cologne/Bonn 2018,* 1–2. Heidelberg, Propylaeum.
van Sickle, C.E. (1930) The public works of Africa in the reign of Diocletian. *Classical Philology* 25.2, 173–179.
Ward, R.B. (1992) Women in baths. *The Harvard Theological Review* 85.2, 125–147.
Webb, R. (2009) *Ekphrasis, Imagination and Persuasion in Ancient Rhetorical Theory and Practice.* Farnham, Ashgate.
Whittow, M. (1990) Ruling the late Roman and early Byzantine city: a continuous history. *Past & Present* 129.1, 3–21.
Yegül, F. (1992) *Baths and Bathing in Classical Antiquity.* Cambridge, MIT Press.
Yegül, F. (2014) Roman imperial baths and thermae. In R. Ulrich and C. Quenemoen (eds) *A Companion to Roman Architecture,* 299–323. Malden, Blackwell.
Yegül, F., Bolgil, M. and Foss, C. (1986) *The Bath-Gymnasium Complex at Sardis.* Cambridge, Harvard University Press.

Chapter 12

Reused columns in an ancient circular economy

Jon M. Frey

Introduction

It has been nearly 25 years since Dale Kinney declared the study of *spolia* to be a 'growth field of art history' (Kinney 1997, 117), and in the time since a steady flow of new books and articles shows that a lively discussion of the semantic possibilities of reuse continues unabated. At the same time, for all its progress in interpreting spoliation as something more than an indication of artistic and social decline, research into the practice remains at the stage of discovery, with the presentation of new examples from an ever-widening range of places and times outnumbering efforts at synthetic analysis (Schattner and Fernández 2009; Altekamp *et al.* 2013; 2017; Frey 2015a; Jevtić and Yalman 2018; Ng and Swetnam-Burland 2018; Underwood 2019; Duckworth and Wilson 2020). Thus, even after two decades, the words of Maria Fabricius Hansen (2003, 14–15) still ring true: 'Extensive ground has been gained regarding a descriptive architectural history of individual buildings. But compared to the investigations into the *whens* and *wheres* of the structures, the *whys* have attracted less scholarly attention.' To be sure there have been attempts at theorising the ancient practice – Richard Brilliant's (1982) classification of *spolia* types, augmented by Anthony Cutler (1999) and Paolo Liverani (2011), or Sarah Rous' (2019) development of the concept of upcycling stand out as good examples – but the field seems to seek entropy more than cohesion. As some like Stephan Altekamp (2013, 159–166) have suggested, in a world that tended toward modular construction in durable materials, the practice of reuse was inevitable and so widespread that asking why people used *spolia* might be akin to asking why people built monumental architecture at all. This does not mean that attempts to gain new insight are pointless. On the contrary, it is in moments when we are challenged to utilise new frameworks – in this case, the increasingly popular discussion of circular economies – and when the ancient and modern worlds are set

Fig. 12.1 Reused columns in the west porch of the Church of Ag. Georgios in Thessaloniki. Note the use of an inverted capital as a column base (Image by Jon M. Frey).

side by side, that we stand to learn the most from the evidence we have already gathered.

Ancient reuse and modern circular economies

At face value, the study of spoliation in antiquity and that of circular economies in contemporary times would seem to have a lot in common. Indeed, a handful of articles suggest that earlier reuse practices might have an important role to play in the development of recycling strategies that replace the linear model of resource extraction, utilisation and discard (Kuijpers 2019; Charlton 2020; Aggeri 2021). There are also methodological similarities, especially when it comes to the difficulty of defining the meanings and limits of the materials and practices defined by the terms '*spolia*' and 'circular economy.' In the decades since Friedrich Wilhelm Deichmann (1975) and Arnold Esch (1969) composed their seminal works on *spolia*, scholars have debated the universality and inclusivity of the concept. All recognise that the term was adapted in the 16th century from the Latin word for 'spoils of war' to describe the fragments of architecture that were incorporated in the buildings of the Christian era. But in seeking to define the concept more clearly, some have restricted the concept to instances of 'meaningful reuse' – a slippery expression that assumes that our perception of the world matches that of the ancient viewer – and limited their studies to a restrictive collection of architectural, sculptural and epigraphic examples beginning in the later Roman period and moving into the Middle Ages (Deichmann 1975; Greenhalgh 2011). Others have pushed these chronological and typological boundaries by studying instances of reuse that range from the Classical period to the modern era, and by placing more traditional examples like the Arch of Constantine or early Christian basilicas along a continuum that stretches from the purely practical utilisation of recycled building materials to the creation of newly carved materials that recall earlier architectural traditions (Esch 1969; Brandenburg 1996; Cutler 1999; Ward-Perkins 1999; Brilliant and Kinney 2011; Frey 2015b; Rous 2019).

This same difficulty with defining the boundaries of the field is also found within the study of the circular economy. As with *spolia*, the origin of the expression 'circular economy' is not entirely clear (Qi *et al.* 2016; Murray *et al.* 2017; Awan *et al.* 2020). Many scholars would point to the 18th century while there are a few who suggest precedents for the practice in much earlier times (Antikainen *et al.* 2018; Mendes 2020). Similarly, the expression itself defies consensus. Recently, Julian Kirchherr *et al.* (2017) assembled 114 definitions only to conclude that the concept means different things to different stakeholders. Their composite definition though seems to capture well enough the subject at hand.

> A circular economy describes an economic system that is based on business models which replace the 'end-of-life' concept with reducing, alternatively reusing, recycling and recovering materials in production/distribution and consumption processes, thus operating at the micro level (products, companies, consumers), meso level (eco-industrial parks) and macro level (city, region, nation and beyond), with the aim to accomplish sustainable development, which implies creating environmental quality, economic prosperity and social equity, to the benefit of current and future generations (Kirchherr *et al.* 2017, 224–225).

Unfortunately, here is where the potential utility of the comparison begins to falter. For while some Greek philosophers like Anaxagoras (fragment 6) and Empedocles (fragment 17) do occasionally express ecologically enlightened ideas about the oneness of nature or, like Plato (*Critias* 111-b) or Pliny the Elder (*Natural History* 33.1.2), even demonstrate an awareness of the destructive human impact on the environment, one is hard pressed to find a call to action similar to the publications of the *Ellen MacArthur Foundation* (https://ellenmacarthurfoundation.org/) or the *Circular Economy Promotion Law of the People's Republic of China* (http://www.fdi.gov.cn/) (On ancient ideas about the natural world, see Hughes 1994). Rather, it would seem that this modern concept of the circular economy is tied too closely to the traditions of industrialised capitalistic society, and even more specifically, to the aggressive and damaging use of petrochemicals over the last century, to find a parallel in the ancient world. As a brief aside, it may be worth pointing out that another mainstay of contemporary economic thought, namely that efficiency is an unquestionable good, may not have been universally accepted in antiquity. To cite just one well known example, Suetonius (*Vespasian* 18) tells us that the emperor Vespasian rejected the unsolicited proposal of a more efficient method of transporting columns through the city with the justification that he needed to provide for the poor, presumably through payment for labour.

Instead, perhaps the greatest contribution that the contemporary conversation around circular economies can make to the study of antiquity is the realisation of just how deeply ingrained in our world view is the linear model of production, use and discard. For the past few centuries and especially since the industrial revolution, the exploitation of people and places for resources to feed ever more complicated linear systems of consumption has become so normalised that this way of seeing the world imperceptibly seeps into our interpretations of evidence from antiquity. Arguably, this explains why *spolia* as an arbitrarily designated category of reuse was invented in the Renaissance but does not become a topic of sustained academic discussion until the middle of the 20th century. It is certainly interesting to note how many studies of *spolia* in the last century begin by referencing the ubiquity and diversity of architectural reuse practices in antiquity (from stones burned in kilns for lime to recut veneers to the conversion of entire buildings) but then end up treating spoliation as somehow a unique class of object (Deichmann 1940, 114–119; 1975, 3–5; Wohl 2001, 85–86; Papalexandrou 2003, 57; Kinney 2006, 233–234. On the anachronism of defining *spolia*, see Kinney 1997, 117–120). Could it be that the field of *spolia* studies is actually motivated by an under-examined industrialist worldview that sees beautiful heirlooms or objects made from rare materials as the only things that are worth being passed down in the face of a use and discard model of economic behaviour that now shapes our consumption practices?

The circularity of columns

Such a conclusion begs the question: what happens when we invert the paradigm and work from the assumption that 'reuse, repair, repurpose' and all the other '10R' words associated with the three-phase conceptual model of circular economies are actually the default practices in antiquity and instances of disposal and discard are the exception (Reike *et al.* 2022)? Perhaps the easiest way to address this question is to turn to the reuse of columns in Roman and Byzantine architecture. More than any other form of masonry in the ancient built environment, columns, especially those carved from a single piece of colourful stone quarried in distant locations, became popular in the Roman era as a sign of power and distinction. They served a function that was consistently appreciated throughout antiquity, and, more importantly, they are not easily altered to serve any other architectural purpose. They are easily identified in the archaeological record and for this reason are some of the most consistently discussed objects in studies of *spolia*. As a result, while one could say similar things about sculpture or inscriptions, it is with columns that the conversations about ancient reuse and calls for a move toward circular economies draw nearest to one another.

According to the available evidence, the Roman use of marble columns went hand in hand with the practice of reuse in cut stone architecture. Where public architecture in the earlier Etruscan tradition relied upon wood and brick walls atop local stone foundations, by the late Republic, prominent Romans began to sponsor the construction of temples and porticoes equipped with large numbers of columns in materials and styles that showed their foreign origins (Beck 2022). This change, which served as a form of social distinction, can be seen most clearly in the group of manubial temples and other monuments in the Roman Forum Boarium. For example, the 1st-century AD natural philosopher Pliny the Elder (*Natural History* 36.24) reports that while serving as aedile in 58 BC, Marcus Scaurus erected a three-storey temporary theatre containing no fewer than 360 columns, the lowest level of which were made from marble and interspersed with 3000 bronze statues. While Pliny does not specifically say so, it is reasonable to conclude that such a large number of columns and statues must have contained recycled elements, probably taken as spoils from various places in the Mediterranean. Peter de Staebler (2018) suggests that local shrines and decorated tombs in the East were a likely source of material. But, even if that was not the case and these were quarried anew, Pliny also reports that a month later, upon demolition of the theatre, Scaurus had the largest of these columns, which measured close to 38 feet (11.6 m) tall, placed in the atrium of his home. Barbara Burrell (2006, 451) argues that after the downfall and exile of Scaurus, these columns were then used in the construction of the Theatre of Marcellus, while Alessandro Viscogliosi (1996, 61–63) reports that others ended up in the Temple of Apollo Sosianus. Thus, one of our earliest examples of the introduction of marble columns into Roman architecture involves at least two phases of reuse. Moreover, what appears to have caught Pliny's attention is not that the columns were reused, but that they were placed in Scaurus' own home after having served a public function.

In another passage, Pliny (*Natural History* 36.13) is even more explicit in recording that in 86 BC Sulla removed columns from the unfinished Temple of Olympian Zeus in Athens for use in temples on the Capitoline in Rome. It is not clear which elements were brought to Rome, but some have suggested that the columnar fragments discovered in a finished state among the remains of the Mahdia shipwreck are to be connected with this episode (Abramson 1974, 9, n. 41). Even if this connection cannot be confirmed, the number and diversity of finished columns recovered from the Mahdia shipwreck alone is impressive (Fig. 12.2; Ferchiou 1994; Hesburg 1994). To be sure this is but one example and shipwrecks containing similar cargoes make up only a small proportion of the total that are known. But while earlier studies of the marble trade focused on the long distance transport of materials originating in imperially controlled quarries, more recent work has revealed a notable number of smaller ships following shorter distance routes and containing a lesser number of carved columns and capitals (Meijer 2002, 148–153; Russell 2011; 2012; 2013; Rice 2016) For example, there is the Şile shipwreck, just northeast of Istanbul, which contained

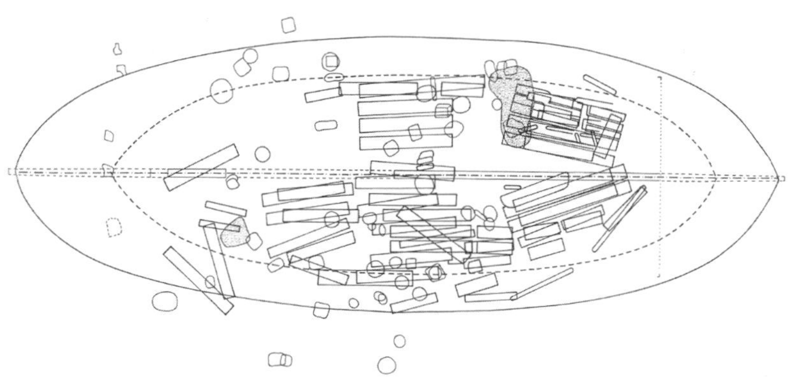

Fig. 12.2 Plan of Mahdia shipwreck showing locations of monolithic columns (Image courtesy of Henner von Hesberg).

two 3.5 m high breccia columns in a finished state along with five unfinished Ionic capitals (Beykan 1988), the 'shipwreck of columns' near the village of Methone in Greece, which is said to have held at least 16 red granite columns taken from a single destroyed or demolished building (Throckmorton and Bullitt 1963), and the Margarina shipwreck that contained at least 20 finished white marble columns (Jurišić 2000, 40). Off the coast of Israel, the later Roman Hahotrim shipwreck has yielded columns marked with crosses indicative of prior use in a church (Kingsley 2004, 60–61).

Then again, since transport by sea was a difficult and dangerous venture, we should probably expect to see an even more active market in recycled architectural materials on a local scale. For this, we can turn to the evidence of marble stockpiles which, since J. Clayton Fant's study of the evidence for the stone deposit at Portus, has become a topic of some discussion (Fant 2001; Russell 2013, 234–239). And while it is likely that stoneyards in major urban centres contained a number of overstock elements, rejects and pieces broken in transport and waiting for other uses, Simon Barker's (2011; 2012) research into Roman demolition practices points to the likelihood that recycled architectural elements are also present in good number. Admittedly, columns form a small proportion of the objects in these stockpiles, but it may well be the case that their particular shape and size rendered them more desirable or, at the very least, less suitable for reworking into other shapes. In this respect, it is worth mentioning Arietta Papaconstantinou's (2012) presentation of a papyrus document from Oxyrhynchus that provides an inventory of columns with their location, dimensions, material and finish. Birgitta Wohl notes the existence of similar drawings during the Renaissance, which were 'most likely done as an inventory for future reuse' (Wohl 2001, 93; Fig. 12.3).

Extant copies of Charlemagne's communication with other elites in Europe suggest that the stockpiling and trading of columns continued into the Carolingian era. One letter from Pope Hadrian to Charlemagne gives him permission to take material from a palace in Ravenna, while another letter from Charlemagne to King Offa of Mercia responds positively to a request for 'black stones' of a determined length. D.P.S. Peacock (1997) suggests that these are columns of Egyptian black porphyry, examples of which have been identified among the architectural remains recovered in Aachen. A scientific analysis of the columns ringing the central rotunda of the Palace Church at Aachen shows that they are from the Mediterranean region and likely come from Ravenna. It is all too fitting that such analysis was possible in part because some of the columns from the Palace Church were taken to Paris as a result of Napoleon's capture of Aachen in 1794 and are now displayed in two different galleries of the Louvre. This fact, as well as the restoration of the majority of the seized columns in 1815 and the reconstruction of the Aachen chapel in the 1840s, suggests that columns continue to be considered important tools in communicating a wide variety of messages. (Konnegen 2007; Shaffer 2012; 2015).

Admittedly these examples lack chronological or spatial consistency, but when taken as a whole, they suggest a conclusion that, as John Onians (1988, 70) has stated,

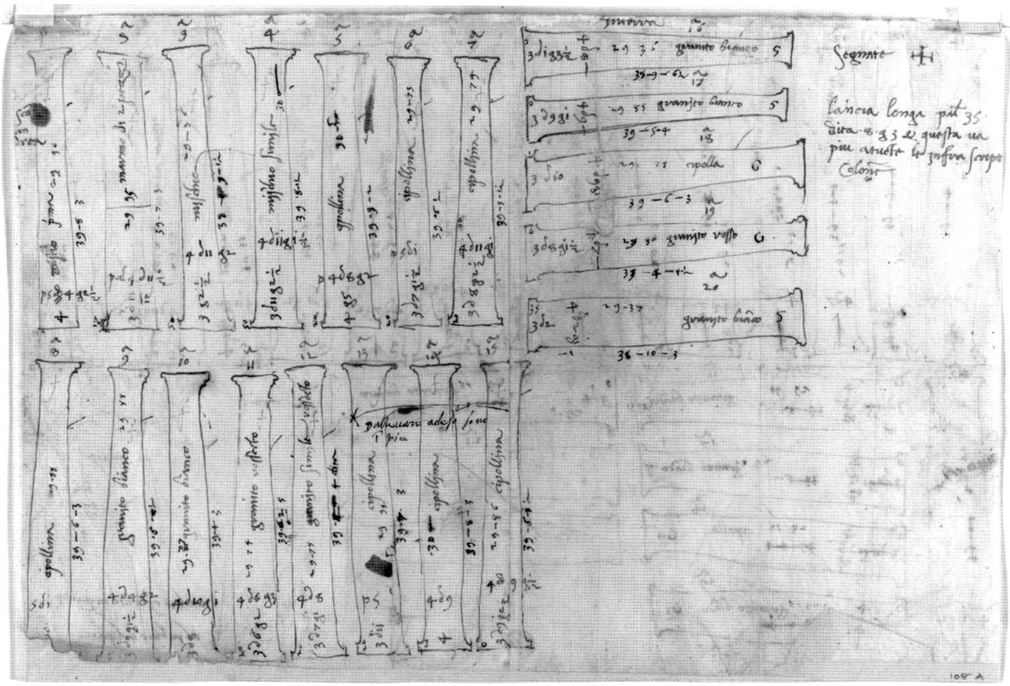

Fig. 12.3 Inventory of columns from Old St. Peter's basilica as recorded by Baldassarre Peruzzi c. 1520. (Image courtesy of the Uffizi Gallery).

'Columns had long been the measure of wealth and status from the time when they were taxed in the Late Republic and the mass production of monolithic columns of standard dimensions in the empire had ensured that they were gradually usable as spolia when architects...began to plunder earlier structures.' Perhaps not too surprisingly then, spoliate columns, specifically those used as vertical supports in the colonnades of mosques and churches, have played a consistently central role in our modern attempts to understand the potential motivations and meanings behind the phenomenon of reuse. In Kinney's (2001, 140–141) summation, 'When we speak of architectural *spolia* we are speaking mostly of the classical colonnade: column shaft, base, capital, and entablature.'

In his landmark study, F.W. Deichmann worked from the arrangement of columns in symmetrical pairs on the façades of sarcophagi to argue that the varied reuse of columns in the naves and aisles of early Christian basilicas were not rule-less but actually followed a sophisticated system that visually marked divisions within these newly ordered religious spaces – his 'Paar-Prinzip' (Deichmann 1940, 122). A particularly convincing example of this was later shown to function in the Lateran Baptistry, in which the pairs of columns and capitals are thought to mark one's progress through the stages of Christian rituals (Onians 1988, 62–63). Since those early stages of study, the variety of potential interpretations has only grown.

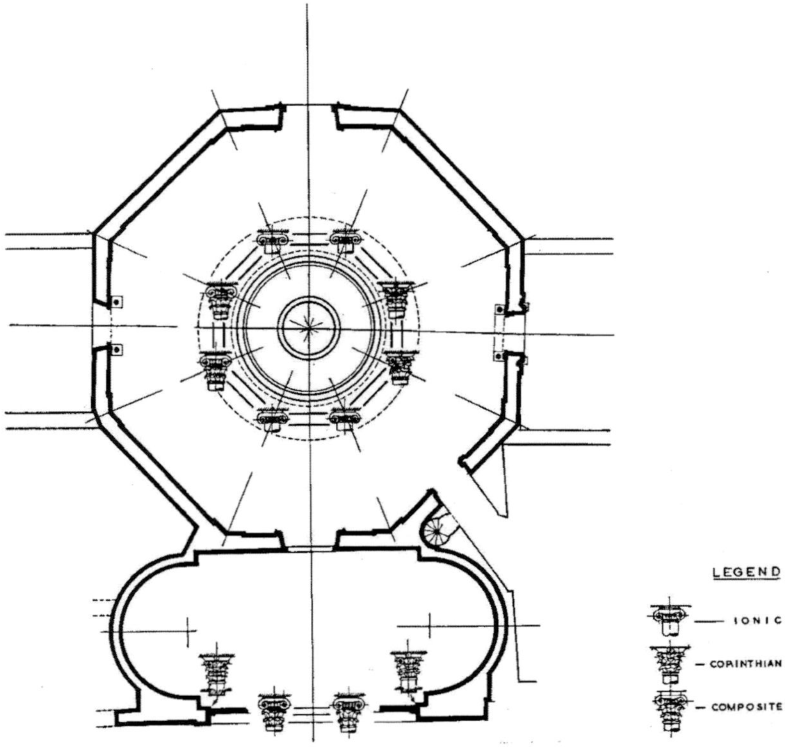

Fig. 12.4 Schematic plan of the arrangement of reused capitals in the Lateran Baptistery at Rome (Image by Don Johnson, provided courtesy of John Onians).

Some have seen in reused columns a desire for political legitimacy (Brenk 1987), while the 'misuse' of columns – more commonly the inverted use of capitals as column bases – has been explained as evidence of the 'victory of Christianity' over paganism (Saradi 1997: Fig. 12.1). Some have taken a more reserved approach by arguing that the reuse of mismatched sets of columns is due to scarcity of a desired architectural element (Pensabene 1997), while others have pointed to the use of newly carved materials of different shapes and styles next to reused columns as proof of the development of an aesthetic of '*varietas*' that becomes popular at this time (Brandenburg 1996; 2011; Brenk 1996). Whether spoliation leads to the appreciation of variety or vice versa remains a point of discussion, but in the end, it is clear that a great deal of the conversation continues to focus on the use of columns and capitals, typically in religious settings to the exclusion of other types of reuse that help to round out the picture.

Some linear examples

Yet, in spite of all the attention they have received in studies in the past, there is still value in examining columns for the insight they can offer into the motivations behind the use of *spolia* more generally. This is because their circularity, in a physical sense, sets them apart, especially when they have been built into the walls of buildings and fortifications and not used in keeping with their original function. After all, most elements of buildings and monuments are rectilinear (ashlars, statue bases, inscriptions) or can be made so with relative ease (theatre seats, funerary stelae, cornices) and are therefore more amenable to repurposing into construction material for the walls of fortifications and other Byzantine and medieval structures. But columns present a bigger building challenge as their shape renders them unsuitable for most reuse scenarios beyond their originally intended purpose. As monoliths, columns are often too long to be used horizontally in walls, and their smoothly finished surfaces make it difficult to lift and manoeuvre them into place. Even as individual drums or cut down into smaller sections, their curvilinear profile leaves significant gaps between elements that must be patched with smaller stones or mortar. If, as has been argued elsewhere, the mismatched dimensions of reused blocks make it even more difficult to build, then columns are the ultimate proof. Understanding this makes it that much more remarkable when we identify columns in the walls of later structures.

Perhaps the earliest and best demonstration of this is an example of reuse that is so historically isolated that its relevance to the larger study of spoliation might be reasonably questioned, but in the context of the discussion of linearity in a circular

Fig. 12.5 North wall of the acropolis at Athens showing the reuse of column drums (Image by George Koronaios, reproduced under CC BY 4.0 license).

economy, it is especially useful as a paradigmatic example. In the aftermath of the 5th-century BC invasion and sack of their city, the Athenians set about clearing and rebuilding their acropolis (Fig. 12.5). In addition to burying many of the dedications that had been erected there over the past century, they also entombed the remains of destroyed buildings in the fortifications that faced north onto the public marketplace below (Tschira 1940; Korres 2002). Scholars have argued that the placement of the Doric entablature blocks in this wall points to an attempt to commemorate a lost building in the walls of the acropolis (Rous 2019, 36–45). More interesting in the present context is the decision to embed an even greater number of unfinished column drums in the same wall. It is readily apparent that their cylindrical shape makes them particularly unsuited to the task of fortifying the acropolis. They are placed in their original orientation, which allows for level horizontal seams, but there are clearly gaps between drums where the thickness of the wall drops to almost nothing. In addition, the placement of the column drums does not follow the standard building practice of staggering vertical joints between blocks, and the resumption of a more traditional ashlar wall leaves a thick ledge at the top of the columns. Even more interestingly, these are not finished drums. Their flutes have not been carved yet, which would suggest that a conscious decision has been made here against the adaptation of architectural members that could have been made available for reuse in keeping with their original function. To use the words of the contemporary discussion on circularity, a linear use and discard model has been chosen over a cycle of reuse here. One could say that the columns have been repurposed, but as with the damaged Archaic statues buried on the acropolis above, any chance of additional reuse according to their original function has been taken away through their burial in a fortification wall (Lindenlauf 1997; Kousser 2009).

The north acropolis wall is quite unique, as more widespread examples of the reuse of columns as building material in fortifications do not appear until around the 5th century AD, when changes in the settlement patterns, combined with a declining sense of security, encourage the construction of a large number of civic and regional defensive works. For the most part, the reuse of columns in such fortifications is utilitarian in nature. There are numerous examples of columns used in the lowest few courses of walls where they seem to have functioned as headers binding the faces of a wall together. There is also the practical matter of columns being more easily rolled into place at the bottom of a wall than lifted into place as with the many other ashlars that appear in higher levels. Occasionally too, columns are found in a secondary position as posts or lintels in gates, but for the most part they are a rare sight in fortification walls.

This is what makes the next example that much more remarkable. In one particularly well-preserved section of the later Roman fortifications at Sparta in the Peloponnese, we find a very unique use of columns (Fig. 12.6; Frey 2015a, 85–127). This fortification starts out like most others, with a heavy use of spoliated ashlars at the lowest courses, but nearer to the top, some columns of different diameters and with different surface treatments are prominently displayed as part of an alternating

Fig. 12.6 Section R of the fortifications at Sparta (Image courtesy of the British School at Athens.)

pattern of rounded and square shapes that has been described as an imitation of a Doric-style frieze course (Gregory 1982, 18). These columns do not appear on the rear face of the wall, where later repairs block our view, but given that the wall is only 2.5 m thick, it does not seem likely that they are preserved to their original height and are somehow being used as headers to bind the faces of the wall around a mortared core. The goal here seems to have been to decorate the face of the wall.

At a height of some 5 m above the foundations, the creation of this layer of blocks would not have been an easy task. Fragments would need to be sorted on the ground, lifted into place in an orderly fashion and then 'patched' so that the inevitable gaps between the curved and angular shapes did not affect the construction of superimposed layers. It is hard to imagine that this type of reuse is driven by necessity or convenience. Moreover, the fact that this is a fortification argues against the types of religious interpretations that are typically associated with 'misused' columns.

But this type of reuse of columns can also occasionally be found in religious settings, so it is worth examining just one of the types of churches that have tended to dominate the discussion of *spolia* over the past half century, the Church of the Virgin of Skripou at Orchomenos (Papalexandrou 1998; 2003, 63–68). Thanks to the fact that

Fig. 12.7 Western façade of the Church of the Virgin of Skripou at Orchomenos, showing the extensive use of column drums in the construction of the upper walls (Image by Jon M. Frey).

the founder's inscription is still embedded in the northwest corner of the structure, it is possible to date the construction of the church to the 9th century. The building has suffered collapse over the centuries and was extensively repaired in 18th century, but Amy Papalexandrou, who has studied them in extensive detail, confirms that the spoliate walls on the church exterior retain their original appearance (Fig. 12.7). This church is rich with recycled material – inscriptions, relief sculptures and other marble ashlars – especially in the lowest zone below the dog-tooth brick frieze that rings the building. Much attention has been given to this lowest band of decoration and the potential meanings of individual reused elements in the context of the Middle Byzantine period, but the sections relevant to this discussion are more than 3 m higher, where all four faces of this building show extensive use of a single set of unfluted limestone columns. Given the thickness of the walls and the diameters of the round shapes, it is most likely that these are column drums which have been positioned in such a way that cuttings for dowels and clamps are still visible. The use of these drums is extensive; in some places only a single course of rounded shapes is visible, but on the west and south façades, several rows of columns have been stacked one atop the other. The height at which these columns have been placed in the wall and the way in which they have been differentiated from the rest of the masonry again argue in favour of an intentional effort that is not based in simple economy. These columns are being reused emphatically and in a way that does not respect their original function. To be fair, Papalexandrou (1998, 103, 252) has suggested on the basis of parallels elsewhere that the walls of this church would have been plastered above the string course, but the exact chronology of this stage in the church's history is not very clear. It is unlikely that it can ever be determined whether the reuse of column drums was hidden from the beginning or whether they were plastered at a later point in time.

Conclusion

In some respects, the visibility of these columns is beside the point. In each of these three examples, columns have been essentially buried in their new contexts so that their former lives as vertical supports in colonnades or decorative façades have clearly come to an end. This 'death' of the column suggests new interpretive possibilities. Mortuary archaeologists often note that death and the rituals that surround it provide groups with a unique opportunity for public expression, whether that be the feeling of loss for a loved one, the sense of retribution at the punishment or destruction of an enemy, or the pride in a family's history, wealth and importance (Williams 2006). They will also point out that, in distinction to contemporary understandings of the practice, cremation in antiquity was typically more costly than inhumation, for the fact that through burning, objects lose all utility for the living (McKinley 2006). There is no opportunity for reuse or recycling. Perhaps the same thing is happening here. By imposing a linear 'end' upon an inherently circular object in an ancient economy that is predisposed toward reuse, ancient builders and architects are seeking to make an especially impactful statement. The statements may differ, and it is up to us to examine the particular details of each instance. But in viewing the spoliation of columns through the lens of an ancient circular economy,

Fig. 12.8 Column in the 11th-century AD Metropolitan Church of Veria, Greece inscribed with a cross (Image by Jon M. Frey).

we might redirect our attention toward an aspect of architectural reuse that has often been ignored in favour of the columnar displays that have always caught the eye of viewers. As compact, durable, uniquely shaped objects, often constructed of exotic materials, columns were and continue to be easily reused as vertical supports. At times they may be remixed and reassembled, but in general they maintain their original function. Such a reality may occasionally enter the minds of builders and architects as they set columns up in their secondary or later phase of reuse, especially in the context of an economic system that, as has been shown at the conference on which this volume is based, tended toward circularity. This realisation may well have motivated some to mark their columns in a way that guarded their then contemporary identity against future reuse and reinterpretation (Kinney 1996; Fig. 12.8). In this respect, it may be worth studying examples of columns inscribed with dedications or official documents. But this reality should also draw our attention even more to instances of spoliation that sought to put a definitive end to the cycle of reuse. It may well be that these unusual instances of discard can tell us even more about attitudes toward the Classical past than the ones that we have traditionally considered remarkable in the context of our own economic world view.

Bibliography

Primary sources

Anaxagoras *The Fragments of Anaxagoras*. Edited by David Sider. Hain, Meisenheim am Glan, 1981.

Empedocles. *Empedocles, the Extant Fragments*. Edited by Maureen R. Wright. New Haven, Yale University Press, 1981.

Plato. *Timaeus and Critias*. Translated by Andrew Gregory. With an Introduction and Notes by Robin Waterfield. (Oxford World's Classics). Oxford, Oxford University Press, 2008.

Pliny the Elder. *Natural History*. Translated by Harris Rackham. Cambridge, Harvard University Press, 1938.

Suetonius. *Vespasian*. Translated by Brian W. Jones, Robert D. Milns. *Suetonius: The Flavian Emperors: A Historical Commentary*. New York, Bloomsbury Academic, 2002.

Secondary sources

Abramson, H. (1974) The Olympieion in Athens and its connections with Rome. *California Studies in Classical Antiquity* 7, 1–25.

Aggeri, G. (2021) From waste to urban mines: a historical perspective on the circular economy. *Field Actions Science Reports. The Journal of Field Actions* Special Issue 23, 10–13.

Altekamp, S. (2013) Architectural re-use processes in Late Antique North Africa – Prolegomena. In S. Altekamp, C. Marcks-Jacobs and P. Seiler (eds) *Perspektiven der Spolienforschung 1. Spoliierung und Transposition*, 159–206. Boston, De Gruyter.

Altekamp, S., Marcks-Jacobs, C. and Seiler, P. (eds) (2013) *Perspektiven der Spolienforschung 1: Spoliierung und Transposition*. Boston, De Gruyter.

Altekamp, S., Marcks-Jacobs, C. and Seiler, P. (eds) (2017) *Perspektiven der Spolienforschung 2: Zentren und Konjunkturen der Spoliierung*. Boston, De Gruyter.

Antikainen, R., Lazarevic, D. and Seppälä, J. (2018) Circular economy: origins and future orientations. In H. Lehmann (ed.) *Factor X: Challenges, Implementation Strategies and Examples for a Sustainable Use of Natural Resources*, 115–129. Cham, Springer International Publishing.

Awan, U., Kanwal, N. and Bhutta, M.K.S. (2020) A literature analysis of definitions for a circular economy. In P. Golinska-Dawson (ed.) *Logistics Operations and Management for Recycling and Reuse*, 19–34. Heidelberg, Springer.

Barker, S.J. (2011) *Demolition, Salvage and Re-Use in the City of Rome 100 BC–AD 315*. Unpublished thesis, University of Oxford.

Barker, S.J. (2012) Roman marble salvaging. In A. Gutiérrez, P. Lapuente and I. Rodà (eds) *Interdisciplinary Studies on Ancient Stone. ASMOSIA IX*, 22–30. Tarragona, Institut Català d'Arqueologia Classica.

Beck, D. (2022) Use, aesthetics and semantics of coloured marble columns in the western Mediterranean during the Late Republic and Early Roman Empire. In A. Haug, A. Hielscher and M.T. Lauritsen (eds) *Materiality in Roman Art and Architecture: Aesthetics, Semantics and Function*, 94–111. Boston, De Gruyter.

Beykan, M. (1988) The marble architectural elements in export-form from the Şile shipwreck. In N. Herz and M. Waelkens (eds) *Classical Marble: Geochemistry, Technology, Trade*, 127–137. Dordrecht, Springer.

Brandenburg, H. (1996) Die verwendung von spolien und originalen werkstücken in der spätantiken architektur. In J. Poeschke, H. Brandenburg and G. Henkel (eds) *Antike Spolien in der Architektur des Mittelalters und der Renaissance*, 11–48. München, Hirmer.

Brandenburg, H. (2011) The use of older elements in the architecture of fourth- and fifth-century Rome: A contribution to the evaluation of spolia. In R. Brilliant and D. Kinney (eds) *Reuse Value: Spolia and Appropriation in Art and Architecture from Constantine to Sherrie Levine*, 53–73. Farnham, Ashgate.

Brenk, B. (1987) Spolia from Constantine to Charlemagne: aesthetics versus ideology. *Dumbarton Oaks Papers* 41, 103.

Brenk, B. (1996) Spolien und ihre Wirkung auf die Ästhetik der Varietas: zum Problem alternierender Kapitelltypen. In J. Poeschke, H. Brandenburg and G. Henkel (eds) *Antike Spolien in der Architektur des Mittelalters und der Renaissance*, 49–92. München, Hirmer.

Brilliant, R. (1982) I Piedistalli Del Giardino Di Boboli: Spolia in Se, Spolia in Re. *Prospettiva* 31, 2–17.

Brilliant, R. and Kinney, D. (eds) (2011) *Reuse Value: Spolia and Appropriation in Art and Architecture from Constantine to Sherrie Levine*. Farnham, Ashgate.

Burrell, B. (2006) False fronts: Separating the aedicular facade from the imperial cult in Roman Asia Minor. *American Journal of Archaeology* 110, 437–469.

Charlton, E. (2020) 3 Ways the ancient world embraced the circular economy. *World Economic Forum*. https://www.weforum.org/agenda/2020/05/circular-economies-ancient-history-recycling/.

Circular Economy Promotion Law of the People's Republic of China. (2023) http://www.fdi.gov.cn/ [accessed 26 February 2023].

Cutler, A. (1999) Reuse or use?: Theoretical and practical attitudes toward objects in the Early Middle Ages. *Settimane di Studio-Centro Italiano di Studi Alto Medioevo* 46, 1055–1080.

De Staebler, P.D. (2018) The re-use of monolithic columns in the invention and persistence of Roman architecture. In D.M. Poljak, K. Marasović and the Association for the Study of Marble and Other Stones Used in Antiquity (eds) *ASMOSIA XI, Interdisciplinary Studies on Ancient Stone, Proceedings of the XI International Conference of ASMOSIA*, 95–100. Split, University of Split.

Deichmann, F.W. (1940) Säule und Ordnung in der frühchristlichen Architektur. *Mitteilungen des Deutschen Archäologischen Instituts, Römische Abteilung* 55, 114–130.

Deichmann, F.W. (1975) *Die Spolien in der spätantiken Architektur: Vorgetragen am 5.7.1974*. München, Beck in Komm.

Duckworth, C.N. and Wilson A. (eds) (2020) *Recycling and Reuse in the Roman Economy*. Oxford, Oxford University Press.

Ellen MacArthur Foundation. (2023) https://ellenmacarthurfoundation.org/ [accessed, 26 February 2023].
Esch, A. (1969) Spolien: zur Wiederverwendung antiker Baustücke und Skulpturen im mittelalterlichen Italien. *Archiv Für Kulturgeschichte* 51, 1–64.
Fabricius Hansen, M. (2003) *The Eloquence of Appropriation: Prolegomena to an Understanding of Spolia in Early Christian Rome.* Roma, L'Erma di Bretschneider.
Fant, J.C. (2001) Rome's marble yards. *Journal of Roman Archaeology* 14, 167–198.
Ferchiou, N. (1994) Recherches sur les éléments architecturaux. In G. Hellenkemper Salies, H.-H. von Prittwitz und Graffon and G. Bauchhenß (eds) *Das Wrack. Der antike Schiffsfund von Mahdia*, 195–208. Köln, Rheinland-Verlag.
Frey, J.M. (2015a) *Spolia in Fortifications and the Common Builder in Late Antiquity.* Leiden, Brill.
Frey, J.M. (2015b) The archaic colonnade at ancient Corinth: a case of early Roman spolia. *American Journal of Archaeology* 119, 147–175.
Greenhalgh, M. (2011) Spolia: A definition in ruins. In R. Brilliant and D. Kinney (eds) *Reuse Value: Spolia and Appropriation in Art and Architecture from Constantine to Sherrie Levine*, 75–95. Farnham, Ashgate.
Gregory, T.E. (1982) The fortified cities of Byzantine Greece. *Archaeology* 35, 14–22.
Hesberg, H. von. (1994) Die Architekturteile. In G. Hellenkemper Salies, H.-H. von Prittwitz und Graffon and G. Bauchhenß (eds) *Das Wrack. Der antike Schiffsfund von Mahdia*, 175–194. Köln, Rheinland-Verlag.
Hughes, J.D. (1994) *Pan's Travail: Environmental Problems of the Ancient Greeks and Romans.* Baltimore, Johns Hopkins University Press.
Jevtić, I. and Yalman, S. (eds) (2018) *Spolia Reincarnated: Afterlives of Objects, Materials and Spaces in Anatolia from Antiquity to the Ottoman Era.* Istanbul, Koç University Research Center for Anatolian Civilizations.
Jurišić, M. (2000) *Ancient Shipwrecks of the Adriatic: Maritime Transport during the First and Second Centuries AD.* Oxford, British Archaeological Reports.
Kingsley, S.A. (2004) *Shipwreck Archaeology of the Holy Land: Processes and Parameters.* London, Duckworth.
Kinney, D. (1996) Making mute stones speak: reading columns in S. Nicola in Carcere and S. Maria in Aracoeli. In C.L. Striker and J.S. Ackerman (eds) *Architectural Studies in Memory of Richard Krautheimer*, 83–86. Mainz am Rhein, P. von Zabern.
Kinney, D. (1997) Spolia. Damnatio and Renovatio Memoriae. *Memoirs of the American Academy in Rome* 42, 117–148.
Kinney, D. (2001) Roman architectural spolia. *Proceedings of the American Philosophical Society* 145, 138–161.
Kinney, D. (2006) The concept of 'spolia.' In C. Rudolph (ed.) *A Companion to Medieval Art*, 233–252. Malden, Blackwell.
Kirchherr, J., Reike, D. and Hekkert, M. (2017) Conceptualizing the circular economy: an analysis of 114 definitions. *Resources, Conservation and Recycling* 127, 221–232.
Konnegen, L. (2007) 'Zu Aach hab ich gesehen die [...] seulen [...] von porfit grün und rot': Die antiken Säulen des Aachener Domes und ihr Schicksal im 18. und 19. Jahrhundert. *Das Münster* 60, 40–50.
Korres, M. (2002) On the north Acropolis wall. In M. Stamatopoulou and M. Yeroulanou (eds) *Excavating Classical Culture: Recent Archaeological Discoveries in Greece*, 179–186. Oxford, Archeopress.
Kousser, R. (2009) Destruction and memory on the Athenian Acropolis. *The Art Bulletin* 91, 263–282.
Kuijpers, M. (2019) Circular economy: ancient populations pioneered the idea of recycling waste. *The Conversation.* https://theconversation.com/circular-economy-ancient-populations-pioneered-the-idea-of-recycling-waste-107332 [accessed 26 February 2023].

Lindenlauf, A. (1997) Der Perserschutt auf der Athener Akropolis. In W. Hoepfner (ed.) *Kult und Kultbauten auf der Akropolis*, International Symposium, 7–9 July 1995, Berlin, 45–115. Berlin, Archäologisches Seminar der Freien Universität Berlin.

Liverani, P. (2011) Reading spolia in Late Antiquity and contemporary perception. In R. Brilliant and D. Kinney (eds) *Reuse Value: Spolia and Appropriation in Art and Architecture from Constantine to Sherrie Levine*, 33–51. Farnham, Ashgate.

McKinley, J. (2006) Cremation ... the cheap option? In R. Gowland and C. Knüsel (eds) *The Social Archaeology of Funerary Remains*, 81–88. Oxford, Oxbow Books.

Meijer, F. (2002) Wrecks in the Mediterranean as evidence of economic activity in the Roman Empire. In W. Jongman and M. Kleijwegt (eds) *After the Past. Essays in Ancient History in Honour of H. W. Pleket*, 135–156. Leiden, Brill.

Mendes, I. (2020) *The Circular Economy an Ancient Term that Became Polysemic*. Lisbon, Lisbon School of Economics and Management Department of Economics Universidade de Lisboa.

Murray, A., Skene, K. and Haynes, K. (2017) The circular economy: An interdisciplinary exploration of the concept and application in a global context. *Journal of Business Ethics* 140, 369–380.

Ng, D. Y. and Swetnam-Burland, M. (eds) (2018) *Reuse and Renovation in Roman Material Culture: Functions, Aesthetics, Interpretations*. Cambridge, Cambridge University Press.

Onians, J. (1988) *Bearers of Meaning: The Classical Orders in Antiquity the Middle Ages and the Renaissance*. Princeton, Princeton University Press.

Papaconstantinou, A. (2012) A Fourth-century inventory of columns and the late Roman building industry. In R. Ast, H. Cuvigny, T. Hickey and J. Lougovaya (eds) *Papyrological Texts in Honor of Roger S. Bagnall*, 215–231. Durham, American Society of Papyrologists.

Papalexandrou, A. (1998) *The Church of the Virgin at Skripou: Architecture, Sculpture, and Inscriptions in Ninth-Century Byzantium*. Unpublished dissertation, Princeton University.

Papalexandrou, A. (2003) Memory tattered and torn: Spolia in the heartland of Byzantine Hellenism. In R. Van Dyke and S.E. Alcock (eds) *Archaeologies of Memory*, 56–80. Malden, Blackwell.

Peacock, D.P.S. (1997) Charlemagne's black stones: the re-use of Roman columns in early medieval Europe. *Antiquity* 71, 709–716.

Pensabene, P. (1997) Amministrazione Dei Marmi e Sistema Distributivo Nel Mondo Romano. In G. Borghini (ed.) *Marmi Antichi, Materiali Della Cultura Artistica*, 43–54. Roma, Edizioni De Luca.

Qi, J., Zhao, J., Li, W., Peng, X., Wu, B. and Wang, H. (2016) Origin and background of circular economy development. In J. Qi, J. Zhao, W. Li, X. Peng, B. Wu and H. Wang (eds) *Development of Circular Economy in China*, 1–19. Research Series on the Chinese Dream and China's Development Path. Singapore, Springer. https://doi.org/10.1007/978-981-10-2466-5_1 [accessed 26 February 2023].

Reike, D., Vermeulen, W.J.V. and Witjes, S. (2022) Conceptualization of circular economy 3.0: Synthesizing the 10R hierarchy of value retention options. In A. Alvarez-Risco, M.A. Rosen, and S. Del-Aguila-Arcentales (eds) *Towards a Circular Economy: Transdisciplinary Approach for Business*, 47–69. Cham, Springer International Publishing.

Rice, C. (2016) Shipwreck cargoes in the western Mediterranean and the organization of Roman maritime trade. *Journal of Roman Archaeology* 29, 165–192.

Rous, S.A. (2019) *Reset in Stone: Memory and Reuse in Ancient Athens*. Madison, The University of Wisconsin Press.

Russell, B. (2011) Lapis transmarinus: Stone-carrying ships and the maritime distribution of stone in the Roman empire. In D.J. Robinson and A.I. Wilson (eds) *Maritime Archaeology and Ancient Trade in the Mediterranean. Proceedings of the 2008 OCMA Conference, Madrid. Oxford Centre for Maritime Archaeology* 7, 137–52. Oxford, Oxford Centre for Maritime Archaeology.

Russell, B. (2012) Shipwrecks and stone cargoes: some observations. In A. Gutierrez, P. Lapuente and I. Roda (eds) *Interdisciplinary Studies on Ancient Stone*, 533–539. Tarragona, Institut Català d'Arqueologia Clàssica.

Russell, B. (2013) *The Economics of the Roman Stone Trade*. Oxford, Oxford University Press.
Saradi, H. (1997) The use of ancient spolia in Byzantine monuments: the archaeological and literary evidence. *International Journal of the Classical Tradition* 3/4, 395–423.
Schattner, T.G. and Valdés Fernández, F. (2009) *Spolien im Umkreis der Macht: Akten der Tagung in Toledo vom 21. bis 22. September 2006*. Mainz am Rhein, P. von Zabern.
Shaffer, J.H. (2012) Restoring Charlemagne's chapel: historical consciousness, material culture, and transforming images of Aachen in the 1840s. *Journal of Art Historiography* 7-JS/1.
Shaffer, J.H. (2015) Picking up the pieces of Charlemagne's column screens: the church at Ottmarsheim, the Westbau' of Essen, and the discovery of Aachen's copies. *Marburger Jahrbuch* Für *Kunstwissenschaft* 42, 7–46.
Throckmorton, P. and Bullitt, J.M. (1963) Underwater surveys in Greece. *Expedition Magazine* 5.2, 17–23.
Tschira, A. (1940) Die Unfertigen Säulentrommeln Auf Der Akropolis von Athen. *Jahrbuch des Deutschen Archäologischen Instituts* 55, 242–261.
Underwood, D.R. (2019) *(Re)Using Ruins: Public Building in the Cities of the Late Antique West, A.D. 300-600*. Leiden, Brill.
Viscogliosi, A. (1996) *Il tempio di Apollo in Circo e la formazione del linguaggio architettonico augusteo*. Roma, L'Erma di Bretschneider.
Ward-Perkins, B. (1999) Re-using the architectural legacy of the past, entre idéologie et pragmatism. In G.P. Brogiolo and B.Ward-Perkins (eds) *The Idea and Ideal of the Town between Late Antiquity and the Early Middle Ages*, 225–244. Leiden, Brill.
Williams, H. (2006) *Death and Memory in Early Medieval Britain*. Cambridge, Cambridge University Press.
Wohl, B.L. (2001) Constantine's use of spolia. In J. Fleischer, J. Lund and M. Nielsen (eds) *Late Antiquity - Art in Context: Acta Hyperborea 8*, 85–115. Copenhagen, Museum Tusculanum Press.

Chapter 13

Mind the gap: researching reuse practices in Palmyra. The example of reused inscriptions

Julia Steding

Introduction

Ancient Palmyra is famous for its numerous monumental buildings and the tombs that were built in its necropoleis, particularly during the first three centuries AD. While the city was already prosperous in the Hellenistic period (Schmidt-Colinet and al-As'ad 2013), in the Roman period the oasis city, located in the Syrian desert, served as a caravan stop between the Euphrates and the Mediterranean Sea. Due to the intense trade between the East and the West that flowed through the city, Palmyra flourished. Simultaneously, the wealth of the city and its inhabitants grew, resulting in the erection of public and religious monuments. At the same time, funerary buildings became more lavish. By the end of the third century AD, however, the entire region around Palmyra was exposed to military conflict. Palmyra was sacked in AD 272/273 by the Roman emperor Aurelian, after Queen Zenobia's attempt to create a Palmyrene Empire, and the city never returned to its former power (for the most recent overviews of the history of Palmyra, see Sommer 2018; Gawlikowski 2019; Raja 2022). Building activity did not come to a halt though, and just 20 years after the destructive events, in AD 293, a military camp (the so-called Camp of Diocletian) and fortifications were erected around the city, the bathhouse was renovated and the urban circuit was repaired. Life went on in Palmyra with political and societal ups and downs, and the city centre was only abandoned in the 9th century AD (for a comprehensive overview of late antique and early Islamic Palmyra, see Intagliata 2018a, with further references).

This paper focuses on building activities from the Roman period onwards, with an emphasis on the reuse of building materials. Reuse was neither new nor unique to Palmyra, but the site, as it does with many other research questions, provides rich material culture that can be drawn upon to study reuse practices in new buildings and the renovation/modification of existing structures. In particular, inscribed stone objects will be used in this paper to a) explore reuse habits in Palmyra, b) showcase how the study of a single dataset can contribute to the discussion of circular economies and c) expose the gap in the evidence for a critical analysis of the material evidence. While keeping the biases of the dataset in mind, this will lead to an overview of the movement of inscribed stone blocks in the city from the 1st century BC to the 7th century AD.

Reuse in Palmyra

To contextualise the present dataset, a brief overview of research on reuse in Palmyra is necessary. The only comprehensive study of reused materials was undertaken by Emanuele Intagliata (2018b; 2022; 2023) in his analyses of the Sanctuary of Baalshamin, which were based on the excavation records of the French archaeologist Paul Collart and his team from the 1950s and 1960s. As Intagliata rightly notes, most of the reused material was not thoroughly documented, and it was only possible to evaluate reuse practices based on the archival materials from the excavations. Besides reused sculptural remains, 53 fragments with inscriptions could be identified as re-employed in late antique and early Islamic structures (Intagliata 2023, Appendix 1, nos. 28–80). Intagliata successfully studied the movement of materials in(to) the Sanctuary of Baalshamin over time and worked out different building practices and the distance traveled by materials based on their size. He also gave an excellent overview of reuse practices in late-antique and early Islamic Palmyra (Intagliata 2022, 422–427; 2023, 99–100), when 'architecture and sculpture from dismissed building was readily accessible ... and the city had already transformed into a large open quarry' (Intagliata 2022, 423).

A recent study of the very small dataset of recarved funerary sculpture demonstrated that only five loculus reliefs showed certain traces of recarving. Additionally, the back side of one relief was used and two reliefs had additions made to them (Steding 2023). It was concluded that the recarving of funerary portraits was not a regular practice in Palmyra, which stands in contrast to many other contemporaneous sites. The study (Steding 2023) formed part of the conference proceedings *Exchange and Reuse in Palmyra* (Andrade and Raja 2023), in which different case studies focused on, among other things, temples as centres of exchange and the reuse of funerary spaces.

Further studies are needed, though. Discussions of reuse are not common in publications on material from the site with the exception of those by Intagliata on the general ways that the Palmyrenes made use of pre-existing resources for their building activities. While the quarries of the city have been well studied (Schmidt-Colinet 2017; Bessac 2021) and provide plenty of information about the stone supply to Palmyra, the circulation of stones within the city has not been examined in detail.

Dataset: reused inscriptions

This paper will not be able to answer all of the questions related to reuse or give an exhaustive overview, but it will get us one step closer to understanding some of the reuse practices that took place in ancient Palmyra. The paper is based on a dataset of reused stone elements carrying inscriptions (see acknowledgements). All of these objects can be ascribed to the religious or public sphere and have been clearly documented as reused. Funerary inscriptions from inside tombs have not been included.

In total, 1133 public and religious inscriptions have been collected within the project 'Circular economy and urban sustainability in Antiquity' (https://projects.au.dk/circulareconomy), based on the corpora of Palmyrene inscriptions: *Corpus Inscriptionum Semiticarum* (*CIS*), *Palmyrene Aramaic Texts* (*PAT*; Hillers and Cussini 1996) and *Inscriptions Grecques Et Latines De La Syrie* (*IGLS*; Yon 2012). In addition, newer works that were published after these corpora and that concern public monuments and structures were also included (al-As'ad *et al.* 2012; Gawlikowski 2017). Out of all the collected inscriptions, only 242 objects were documented to have been found *in situ*; 233 have an unknown/undocumented context. Thirty-five objects were reused in modern contexts, but as the focus of this paper is the reuse practices of antiquity, reuse in modern buildings has been excluded from the present study. This means that 658 inscriptions are from a known secondary context. Out of this dataset, only 185 stone elements can be identified as reused with certainty (Appendix 1). Even this number must be approached with caution, as there is an obvious gap in the dataset that will be discussed below.

The 185 objects are divided as follow (Fig. 13.1): 30 objects could only be identified as 'stone blocks' and 10 as 'unknown fragments,' with no further specification possible. The largest group of identifiable objects are altars (27 objects), followed by statue consoles (19 objects), column drums (16 objects) and door lintels (12 objects). Ten further objects were identified as plaques. Represented by less than ten objects

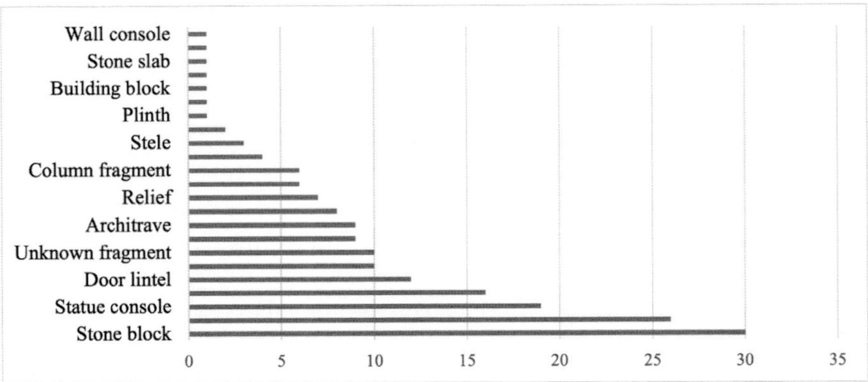

Fig. 13.1 Number of inscribed objects per object type (total: 185).

each are lintels, architraves, unidentified consoles, reliefs, statue bases, column pieces, cornices, stele and tablets. Only one example has been identified of each of the following, a plinth, a pillar, a building block, a crater, a stone slab, a statue and a wall console. The inscriptions on these objects are of different types; honorific (54 objects), votive (53 objects) and funerary inscriptions (37 objects) are the most common, while some beneficent (7 objects), dedicatory (5 objects) and legislative (2 objects) inscriptions are also represented. The inscriptions often mention the name of the individual or group of individuals who had commissioned the inscription and the name of the person or deity being honoured. A small number of the inscriptions include a date, often when a building or parts of a building were erected or dedicated, making is possible to determine a *terminus post quem* for the initial use of the inscribed object. The earliest date mentioned is a dedication of the priests of Bel from AD 44 (year 269) on a statue base (Appendix 1, no. 79), and the latest date mentioned is a dedication to Zeus from AD 302 (year 613) on an altar (Appendix 1, no. 156).

In the public and religious sphere, Palmyrene, Greek, Latin and any of these languages in combination as bilingual and trilingual inscriptions appear. The same picture is reflected in the reused inscriptions: 139 are in Palmyrene, 22 in Palmyrene and Greek, 20 in Greek, two in Latin, one in Greek and Latin, and one in Palmyrene, Greek and Latin. As the language and type of inscription cannot be linked to a specific reuse habit, this information will not be considered in the following sections, but they demonstrate the complexity of the dataset.

Mind the gap

Before any discussion based on the present dataset can take place, it is important to point out its weaknesses. The most obvious gap is that it only includes the known inscribed blocks that come exclusively from the larger structures of the city in and around which most archaeological work was conducted. This leaves us with a spatial gap of unexplored areas of the city where little to nothing about the use and reuse of (building) materials is known. This concerns mainly living quarters as well as industrial areas and unexplored public and religious spaces. The picture thus mainly reflects reuse related to materials from and/or used for specific monuments from the religious, funerary and public spheres. This view is even more biased due to the fact that the Temple of Baalshamin has been studied in terms of reused materials in more detail than anywhere else, as mentioned above; 25% of the present dataset thus originates from a single context, shifting the focus towards one specific context from the religious sphere.

The second gap in the dataset emerges from issues related to documentation. Four hundred and seventy-three objects were found in secondary contexts but have not been properly documented as reused. This leaves us with the question of whether these should be included in the dataset or not. The main difficulty is the description and documentation of the find contexts – or rather the lack thereof. Often it is unclear what the terminology used (*e.g.*, reemployed *versus* reused) implies. Was an object reemployed at its original location, *i.e.* during renovation, or reemployed in a new context in antiquity? Was it reemployed in the modern era during restoration work? Moreover, if images of objects are included in publications, the focus is on the inscriptions, and the photos mainly show the object removed from its find-context. The second gap in the data is thus objects that were excavated from a secondary context, where the circumstances of their reuse is unclear.

In cases where reuse was properly documented, a third issue appears: the lack of information on the exact placement of the stone. In none of the 185 cases of certain reuse was the exact employment of the stone in its new context noted. When reused as filling material, *e.g.*, in a foundation, the exact placement might not matter or give more information, but if an object was reused in a wall, it may provide more information. However, the placement of the blocks was not documented, such as whether the inscription was facing outside or the block was turned upside down. Therefore, it is not possible to say how the reused blocks were exactly used.

The fourth challenge lies in the dating of the objects and the dating of their original use and reuse. If a date is given in the inscription, the first use of the stone in a specific context can be determined. If no date is given, it is possible to approach the date of the original use through its context, *i.e.* the year(s) of construction. For example, an altar without an inscribed date but assuredly associated with a specific temple would not have been created before the construction of the temple, giving a *terminus post quem* for the original use of the altar. The point in time of the reuse is harder to determine, as it is not always clear when or even in what kind of structure the object was placed. This uncertainty mainly derives from the lack of documentation. If something was recorded as reused 'in the Camp of Diocletian', neither the exact location nor the period of reuse can be determined if not specifically mentioned by the excavators. It could be part of the first construction phase in the late 3rd century AD or part of any later construction or renovation in the centuries after. The fourth gap thus concerns the movement of objects over time.

Combined, these lacunae in the data hinder the understanding of the reasons behind these reuse practices. Georgios Pallis (2019, 60), in his publication on the second life of inscriptions in Asia Minor, rightly mentions that information about the exact circumstances of reuse is required if we want to know if there is any symbolic meaning to it. We do not have this information for the present dataset, and this limits our ability to consider the reuse of inscriptions in any symbolic manner. This section has demonstrated the limitations of the dataset and the archaeological evidence, not to criticise previous publications or scholars and their work. The interest of the excavators was clearly the inscriptions and not the objects as building material. This section is also meant to raise awareness of our dependency on the documentation of contexts and the questions that scholars should keep in mind when researching building practices (see also Bavuso *et al.* 2023).

Movement in space

Original contexts

The first aspect to be studied in this paper is the movement of the stone blocks within and between monuments (Fig. 13.2). In some cases, it was possible to determine that an object was first used in a certain building (Fig. 13.3), mainly due to the mention of a deity that had a temple dedicated to them in the inscription or the location of a grave to which a foundation text originally belonged. In total, 98 objects stem from the religious sphere. The largest share belongs to the Temple of Baalshamin (Appendix 1, nos. 1–50, 156, 173), with a total of 52 objects: 11 statue consoles, seven column drums, seven stone blocks, five altars, three architraves, two lintels, two reliefs, two cornices, one statue base, one plinth, one stele, one column and one console. Other temples from which building material originated are the Temple of Bel (11 objects: Appendix 1, nos. 80–84, 157, 161, 168–169, 176, 182), the Temple of Allat (7 objects; Appendix 1, nos. 55–59, 95, 150), the Temple des Enseignes (2 objects; Appendix 1, nos. 93–94) and the Temple of Nabu (1 object; Appendix 1, no. 139). Another 24 objects

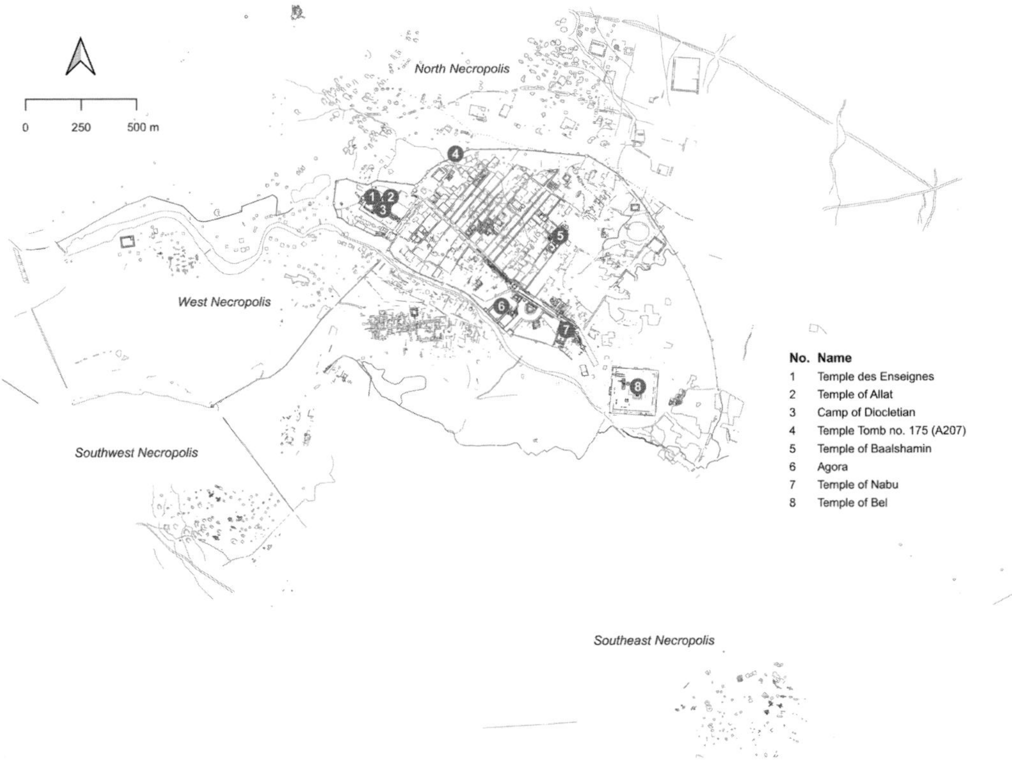

Fig. 13.2 Map of Palmyra with the context in which reuse took place (Map by Katarína Mokránová, courtesy of the Palmyra Portrait Project; Based on extended Topographia Palmyrena (Schnädelbach 2010) data, provided courtesy of Deutsches Archäologisches Institut (Orient-Abteilung, AS Damaskus)).

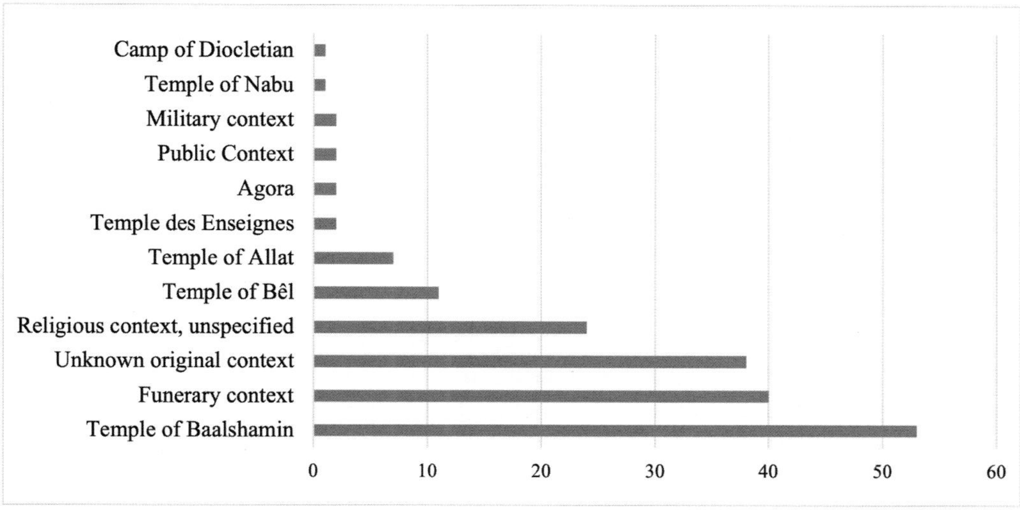

Fig. 13.3 Number of objects from each original context (total: 185).

(Appendix 1, nos. 78, 85, 112–113, 128–132, 143, 146, 158–169, 162–167, 172, 174, 179), based on the character of their inscriptions, came from the religious sphere but could not be allocated to a specific sanctuary.

The second largest original context is the funerary sphere, with 40 objects (Appendix 1, nos. 51–54, 63, 86–87, 96, 101–108, 120–127, 133–138, 140–142, 144–145, 149, 170, 175, 180, 185). In seven cases it was possible to determine that the objects originally belonged to the Northern necropolis (Appendix 1, nos. 121, 134–135, 138, 140–142), and in five cases they came from the Western necropolis (Appendix 1, nos. 107–108, 133, 144–145). The remaining objects could not be assigned to a necropolis, but based on the inscriptions' context, it is evident that they originally were used in the funerary sphere; they could thus have been sourced from either of the mentioned necropoleis or those located in the south of the city. Only in one case was it possible to gain more insight into the original context; a door lintel was originally installed in Temple Tomb no. 175 (A 207) in the North necropolis, informing the reader, in Greek and Palmyrene, that the tomb was built by 'Zenobios and Samuelos, sons of Levi, sons of Iakoubos, sons of Samuelos, in honour of Levi their father, for themselves and for their brothers and for their sons and for their grandchildren and for their descendants, for all time, in the month of Xandikos, the year 523' (Yon 2012, 358; IGLS 471; Appendix 1, no. 134). Three objects were originally installed on the Agora (Appendix 1, nos. 147–148, 177), and one piece belonged to an unknown location in the public sphere (Appendix 1, no. 88). One object can be assigned to the Camp of Diocletian (Appendix 1, no. 111) and two objects to an unspecified military context (Appendix 1, nos. 109–110). A total of 38 objects could not be assigned to any context, but from the inscriptions, object type and/or the location in which they were found it became evident that they must have been moved and reused.

When comparing the number of objects that originated from the different contexts, we see that the religious context is predominant, followed by the funerary context. This is, on the one hand, a reflection of the research focus on the religious and funerary monuments from the city and, on the other hand, due to the nature of the documented inscriptions. In the entire dataset of public inscriptions, the majority do not provide enough information to determine their original use or if they had been moved from their original spot.

New contexts

When looking at the find spots (Fig 13.1) of the reused objects, there are even more contexts (Fig. 13.4). Most objects were found in the religious sphere of the city. Fifty-four objects were reused in the structures of the Temple of Baalshamin (Appendix 1, nos. 1–54), 24 in the Temple of Bel (Appendix 1, nos. 69–92), 14 in the Temple of Allat (Appendix 1, nos. 55–68) and eight in the Temple des Enseignes (Appendix 1, nos. 93–99). The second largest number of objects come from the military context. Nineteen objects were found in the Camp of Diocletian (Appendix 1, nos. 101–120), 18 objects in the city walls (Appendix 1, nos. 120–136) and five in an unspecified part of the rampart (Appendix 1, nos. 137–141). One object was found in an unspecified military context (Appendix 1, no. 142). Only three reused objects were documented in the funerary sphere (Appendix 1, nos. 144–146) and two objects in the Agora (Appendix 1, nos. 147–148).

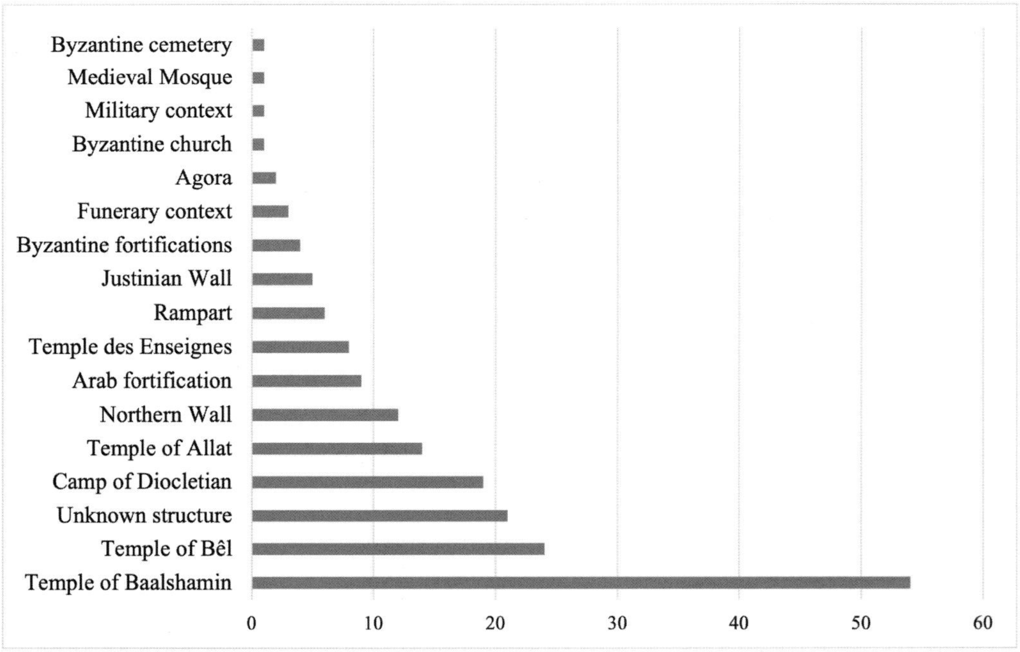

Fig. 13.4 Number of objects from each reused context (total: 185).

In the remaining cases, the exact find context was not specified. Thirteen objects came from unidentified structures (Appendix 1, nos. 172–182,184–185) and one from an unidentified late antique structure (Appendix 1, no. 183). Nine objects relate to Arab fortifications (Appendix 1, nos. 158–160, 162–167) and four to Byzantine fortifications (Appendix 1, nos. 152–55). There are some further contexts with vague descriptions with one to two objects found in each (Byzantine structure, Muslim structure, Byzantine church, medieval structure, Source of Sérail, Arab structure, medieval Mosque, Byzantine cemetery, 12th-century context, gardens).

From the old to the new context

In most cases, the materials circulated within the same complex. Fig. 13.5 shows the relationship between the original context and the reuse context. It does not focus on the number of inscriptions that moved from one context to another but visualises that stones did move from one context to another.

The objects from the Temple of Baalshamin were mainly reused in the temple when it was rebuilt (for a detailed discussion, see Intagliata 2018; 2023; Appendix 1, nos. 1–50). Only one object, an architrave from the temple, was found reused in a wall outside of the temple (Appendix 1, no. 156), and four objects were moved to the temple from the funerary sphere (Appendix 1, nos. 51–54). Objects from the Temple of Allat were mainly reused within the temple as well (Appendix 1, nos. 55–59); only one altar was moved to the Temple des Enseignes (Appendix 1, no. 95) and one column drum into an unidentified Byzantine structure (Appendix 1, no. 150). From the Temple of Bel, five objects out of 11 were reused in the temple (Appendix 1, nos. 80–84), while four objects were moved into

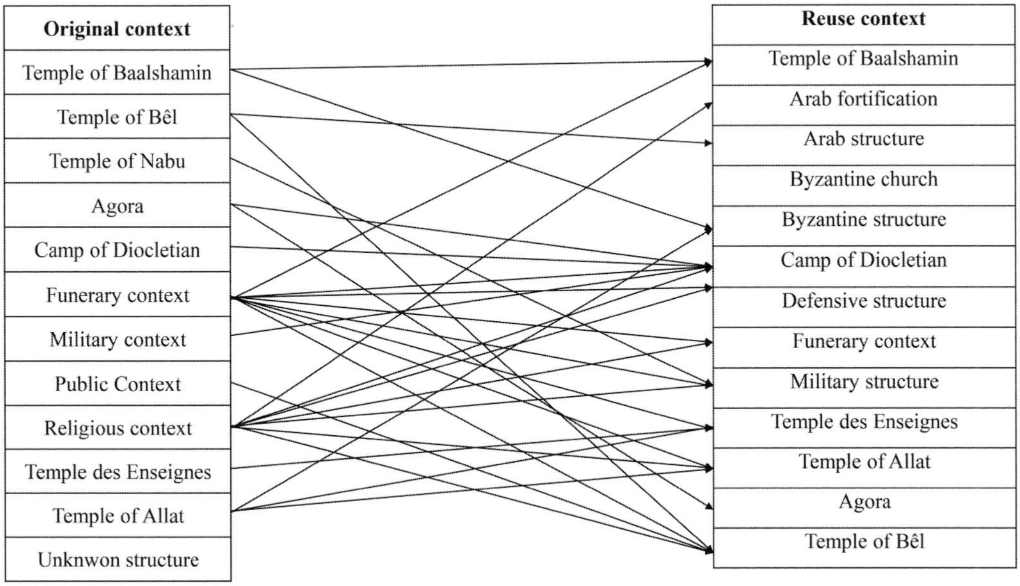

Fig. 13.5 Movement from the original context to the reused context (presence of a connection).

'Arab' contexts (Appendix 1, nos. 157, 161, 168–169) and two to unidentified structures (Appendix 1, nos. 176, 182). Two objects, an altar and a statue, from the Temple des Enseignes were reused in the same temple (Appendix 1, nos. 93–94).

Objects from the funerary sphere, mainly lintels and plaques, were reused in many different monuments and structures around the city. Nineteen objects were reused in different parts of the city wall and ramparts (Appendix 1, nos. 120–127, 133–138, 140–142, 144–145) and eight in the Camp of Diocletian (Appendix 1, nos. 101–107). Four objects were reused in the Temple of Baalshamin (Appendix 1, nos. 51–54), two in the Temple of Bel (Appendix 1, nos. 86–87) and one object each in the Temple of Allat and the Temple des Enseignes (Appendix 1, nos. 63 and 96). Two objects from the funerary sphere were reused in the funerary context (Appendix 1, nos. 144–145).

In summary, the inscribed objects with a known original context mainly moved within the same context, while only a small number moved to other contexts. Within the religious sphere, there seems to be some evidence that objects were moved around when a temple was renovated or rebuilt, as will be discussed below in more detail. However, since the evidence is biased towards the Temple of Baalshamin, it is not possible to argue that the majority of objects were generally reused within the same context. What can be said with more certainty is that a large number of objects were moved into defensive structures, such as the city walls, the ramparts and the Camp of Diocletian, that were erected at the end of the 3rd century AD.

Funerary inscriptions rarely stayed in the same context but moved into other, different contexts frequently. The only two examples that remained in the funerary context are a foundation and a cession inscription from unidentified tombs in the West Necropolis (Appendix 1, nos. 144–145) that were reused in a Byzantine hypogeum at the foot of the Umm el-Qis (Yon 2012, 315–316). The few examples that can be securely connected with the West and North necropoleis were reused in the Camp of Diocletian and the northern part of the rampart (Appendix 1, nos. 107–108, 121, 133–135, 138, 140–142, 144–145). This could mean that distance played a role. The proximity of these two necropoleis to the northern structures of the city was the main reason why the objects were reused there. However, in many other cases it is not clear from which necropolis the objects originated, and it therefore cannot be said for sure that distance always played a role.

Movement in time

Another aspect that can be studied is the period in which the stones were reused. Many of the objects cannot be dated very precisely for either their first use or their reuse, and so the following section is partly speculative. Nonetheless, looking at the objects that can be dated with some certainty helps us to better understand the dynamics of reuse in the city. In total, 63 of the 185 objects have secure dates that tell us their first use and their (last) reuse. Some of these dates are verified by the inscription and knowledge of the exact circumstances of their reuse and some are dated from a verifiable *terminus ante quem* or *terminus post quem*. Even though this is just a fraction of the entire dataset, we can study the chronology of reuse to some extent.

First, a few notes on the dates: the only secure dates for an object's first use are those mentioned in inscriptions (*e.g.*, when a dedication took place). It can be assumed that the object was first in use around the time the inscription was incised and thus in the year that is mentioned in the inscription or shortly thereafter. In the case of most other objects, it can only be said that they most likely had not been used before the construction of the structure in which they originally belonged. For example, tombs were not regularly used after the sack of the city in AD 272/273 (the last founder inscription dates to AD 174; Gawlikowski 1970, 210–211; see also Raja *et al.* 2021; Raja 2022, 86), and so the inscriptions originating from funerary contexts must predate the sack.

When the stone elements were reused is even more challenging to determine. Often, as has been pointed out before, the information given on the find context is sparse. For example, according to its publication in the *Inventaire des inscriptions de Palmyre*, an inscribed column drum (Appendix 1, no. 157) was found in a 'mur arabe' (Cantineau 1933, 45). An exact date is thus difficult to identify, but it can be concluded that the object was reused only after the Muslim conquest of the city in AD 634. In cases where objects were used in foundations, we know that the reuse must have occurred during the building of this later structure.

When studying the evidence with relatively secure dating for the original use and reuse, most objects seem to have been reused after the sack of the city in the late 3rd century AD. However, in one building complex, reused materials were employed much earlier: the Temple of Bel. Eleven inscriptions, mostly from unidentified original structures, can be dated to the Hellenistic period (Appendix 1, nos. 69–79). They were discovered in wall T, a foundation wall between the cella and the propylaea. This was part of the first enclosure of the sanctuary that was finished in AD 32. The reuse of stones within the same context thus occurred before the temple was dedicated in AD 32. At a later point in time, in the 7th century, three inscribed blocks from the 1st to 2nd centuries AD were reused (Appendix 1, nos. 82–84). In the Temple of Baalshamin, six reused inscriptions date from the early 1st to the early 4th centuries AD, and all of them were reused in the Byzantine temple structures (Appendix 1, nos. 27, 35, 43, 45, 50, 53).

The majority of the reuse, however, can be dated to the construction of the military camp and the city walls. Of the securely dated objects, nine, dating from AD 115 to 217, found their way into the Camp of Diocletian and 22, dating from AD 52 to 302, into the first phase of fortifications in the late 3rd and early

4th centuries AD (Appendix 1, nos. 101–131, 137–145). There are a few securely-dated pieces that have been found in later phases of the fortifications (5 objects reused in the mid-5th century and 7 objects reused in the 7th century: Appendix 1, nos. 132–136, 158–160, 162–167), and all of these were also originally inscribed between the mid-1st and mid-3rd centuries AD.

The objects from the funerary sphere date from 12 BC to AD 241, and all of them were reused in the late 3rd and early 4th centuries (Appendix 1, nos. 96, 101–107, 102–121, 123–124, 137–138, 140–142, 144–145). They are further proof of the urgent need for building materials in this period. Some were moved long after their initial use, and some only around half a century after their initial use. What remains unclear is if any reuse happened between the first use of the stone and the final use in the structure from which it was excavated or documented. It is of course possible that stone objects moved around more frequently and that there are multiple phases of reuse that cannot be reconstructed.

Reusing inscribed blocks in Palmyra

When looking at studies on the reuse of inscriptions in antiquity, often a symbolic meaning is assumed for the reuse. Alison Cooley (2000, 7), for example, states that inscriptions on durable material were made to be seen by multiple generations. The reuse of uninscribed stone has also been seen as a symbolic act by some scholars. Elisha Dumser (2018) brings forwards three channels through which meaning can be conveyed: first, through the physical material of an architectural element or the subject matter of a sculpture; second, through the very act of appropriation; and third, through references to the original context of the reused object. The evidence from Palmyra suggests that the stone blocks with inscriptions were not reused because of their inscriptions but that they were seen as building material – and sometimes even just as filling material, as in the 14 proven cases of reused inscriptions in foundations (Appendix 1, nos. 4, 27–28, 58, 63, 85–86, 88, 92–93, 99, 103, 116, 142).

It cannot be ruled out that single objects were placed in a way so that the inscription was visible in the new structure, but there is no mentioning of this in the literature. In some cases, the opposite has been identified; the inscription on a door lintel, removed from a tomb and reused in a hypogeum, was turned towards the ground (*IGLS* 402/*PAT* 1134; for the documentation of the find, see Cantineau 1930, 530). A symbolic function of the inscription can be excluded here. The same was concluded by Pallis for the evidence from Asia Minor. Inscribed blocks were mainly used as raw material, and often the texts were reversed or hidden; and even if the inscription was still visible, it could have been covered with marble revetments, paintings or plaster (Pallis 2019, 67).

The use of inscribed blocks as *spolia* in fortifications is also well attested in many cities in Asia Minor, but slightly later than in the case of the Camp of Diocletian and parts of the city wall in Palmyra. In Aphrodisias, the inscribed parts of some reused stone blocks were systematically hidden. In the Byzantine period in particular, it was common to use epigraphic material in churches (Pallis 2019, 61–64). Both of these reuse practices can also be attested in Palmyra. These objects were used as convenient sources of stone, not as symbolic pieces employed for their original meaning.

An analysis of the discussed evidence alongside the development of the city and its monuments demonstrates different phases of renovation and rebuilding that are reflected in the material evidence. The most can be said about the temples of Bel, Allat and Baalshamin, which underwent massive alterations and were in use over long periods of time. The Temple of Allat was erected in the 1st century BC and underwent multiple building phases until its destruction in AD 272 by Aurelian's troops. It was later restored, and an improvised sanctuary existed until the 380s when it was destroyed. Parts of the temple were used as living quarters for the stationed military commander, and the building was in use even after the military had left Palmyra in the Umayyad period (Gawlikowski 2017, 27–28). The evidence from the reused inscriptions shows that honorific texts, most dating before AD 273, were used as building material for either the restorations or later alterations. The available material mainly came from the pre-existing temple structures, however, an object from the funerary sphere, dating to AD 223, was moved to the temple for use in the foundation at a later period (Ruprechtsberger 1987, 306; Appendix 1, no. 27).

The first place of worship in Palmyra for Baalshamin dates to around AD 20. In AD 67, a large courtyard and porticus were erected, removing any pre-existing structures aside from the northern buildings which remained untouched. There were building activities between AD 90 and 150, but the sanctuary was unaffected by the sack of the city in AD 273. At the end of the 4th century or in the first half of the 5th century AD, the temple was transformed into a church, when the orientation of the building was also changed. Renovations can be dated to the 6th century AD, and the Arab occupation in the 7th century also left its mark on the structures (Collart and Vicari 1969, 89–94). The exact reuse of inscribed blocks has already been demonstrated by Intagliata (2018b; 2022; 2023) and shall not be repeated here. It is just important to highlight in this context that inscribed stones were used for the restorations after AD 273 as well as conversions in later centuries.

The Temple of Bel was the main sanctuary of Palmyra, and the structure was built on some of the oldest remains of the city. It was constructed between AD 19 and 32. In AD 80, a podium was added. During the attack of Aurelian's troops in

AD 273, the temple was looted, but restorations took place just one or two years later. In the 4th century, the temple was transformed into a church, and in the 12th century, the ruins of the temple/church were transformed into a mosque. In the early 14th century, the structure was transformed into a second mosque (Seyrig *et al.* 1975, 160). Inscribed objects from an earlier Hellenistic structure at the same location were first reused when the temple was constructed, and inscribed blocks were also reused later, especially in the 7th century, when the structure was already in use as a church.

The evidence from the Camp of Diocletian and the ramparts around the city, which were built in the late 3rd and early 4th centuries AD (see Intagliata 2017; 2018a, 83–96), reflects the rapid need for building materials when the Legio I Illyricorum was stationed in Palmyra (for the military in Palmyra after AD 273, see Intagliata 2018a, 69, with further references). This is similar to many other cities around the same time (see the volume by Intagliata *et al.* 2020, with specific contributions on sites in Syria by both Hof and Blétry) that had city walls erected. Most of the inscribed and reused objects found their way into city walls, reflecting the urgent need for building materials for defensive structures, particularly in the late 3rd and 4th centuries, as well as in later centuries when there was military tension in the region and natural disaster left its traces on the cityscape (for this period, see Intagliata 2018a, 103).

Here, a brief discussion of the limestone quarries of the city is needed. The Palmyrene building industry was mainly dependent on the local limestone. It was used for structures in the public, religious and funerary spheres. Imported stones were rare and are limited to a few columns of red granite from Aswan (Hirt 2021), the marble decoration (Wielgosz 2013) of the Bath of Diocletian and several marble sculptures from different locations (Wielgosz 2010). Additionally, a few capitals made of Prokonnesian marble and fragments of pavonazzetto marble and possibly cipollino marble were found (Hirt 2021, 101–102). The local limestone did not have to be transported further than 15 km, which is the distance from the city centre to the farthest quarry to the northeast (Schmidt-Colinet 2017, 168). The transport of new limestone to the city was thus neither laborious nor expensive, as was the preference in many cases across the Roman Empire. According to Ben Russel (2013, 17), building blocks were mainly acquired from nearby quarries, ideally transportable to the building site within one day of overland transport.

Towards the middle of the 2nd century AD, the building boom in the city manifested itself in the change of the quarrying techniques around this time, which allowed for more stone to be quarried (Schmidt-Colinet 2017, 189). No evidence of reuse can be detected around this time, indicating that (inscribed) blocks were not used during this building boom. After the sack of the city, however, all available materials from the city itself, whether inscribed or not, were used for building. In the case of changes to an existing structure, *e.g.*, the extension to the Temple of Bel, it made sense to use the available stone and reuse it for the new foundations or additional walls. In other cases, it may have been easier to use already carved blocks from nearby buildings – or even from structures further away. This would have spared the craftsmen the time and labour of quarrying new blocks (Barker 2010). Existing stone could have been adjusted or used unaltered for a new purpose, and the regularity of this practice is reflected in the dataset of inscribed objects. This decision was possibly a consequence of the urgent need for materials in a time of conflict, or it could also reflect a lack of funding for the acquisition of new stone as well as a decline in interest in the existing monuments (or rather their original use/meaning) and the benefactors immortalised in the inscriptions.

Conclusion

As is evident from this paper, the gap we are dealing with in the data about reuse is massive. The first problem is the bias in the available data; only some parts of the city and specific contexts have been excavated properly and only in some cases has the reuse of material been documented and published properly. Often there is a lack of information on the exact manner of reuse and often even on the context in which the object was found. If an altar was found in the city wall, reuse can be assumed even if it is not necessarily mentioned in the literature, due to the the inscriptions themselves being the main focus of the publications. The dating of the original and reuse contexts complicates the analysis even more.

However, the available dataset of 185 objects with inscriptions has showcased that reuse definitely took place during all phases of the city – from the Roman period until the 7th century. Often, the stone was reused within the same context, and the distance between the original placement and the reuse was, in most cases, not very far. Although it should be noted that in some cases, especially with material from the funerary sphere, stones traveled a considerable distance. In his article on reuse within the Temple of Baalshamin, Intagliata (2023) was able to prove that the dimensions of the reused objects could indeed also play a role in the decision taken to reuse a block or not. This methodology could also be employed for the present dataset, but this lays sadly outside the scope of this paper.

Acknowledgements

The dataset in this article based on data collected within the project 'Circular economy and urban sustainability in Antiquity,' headed by Rubina Raja and funded by the Carlsberg Foundation and Augustinus Foundation. I am grateful to Rubina Raja for trusting me with a selection of the dataset for this paper. This work was supported by the Danish National Research Foundation under the grant DNRF119 - Centre of Excellence for Urban Network Evolutions (UrbNet).

Bibliography

Abbreviations

RES = Répertoire d'épigraphie sémitique, 1900–1968, 8 vols. Paris, Imprimerie nationale.
CIS = *Corpus inscriptionum Semiticarum.* Paris, E Reipublicae Typographeo.
PAT = Hillers, D.R. and Cussini, E. (1996) *Palmyrene Aramaic Texts, Publications of the Comprehensive Aramaic Lexicon Project.* Baltimore, John Hopkins University Press.
IGLS = Yon, J.-B. (2012) *Inscriptions grecques et latines de la Syrie 17.1. Palmyre.* Beirut, Institut Français du Proche Orient.

Secondary sources

al-As'ad, K., Gawlikowski, M. and Yon, J.-B. (2012) Aramaic inscriptions in the Palmyra Museum: new acquisitions. *Syria: Archéologie, Art et Histoire* 89, 163–183.
Andrade, N. and Raja, R. (eds) (2023) *Exchange and Reuse in Roman Palmyra Examining Economy and Circularity.* Turnhout, Brepols.
Barker, S. (2010) Roman builders – pillagers or salvagers? The economics of deconstruction and reuse. In S. Camporeale, H. Desalles and A. Pizzo (eds) *Arqueología da la construccíon II. Los procesos constructivos en el mundo romano: Italia y provincias orienteles, Certosa di Pontignano, Siena, 13-15 de noviembre de 2008,* 127–142. Madrid/Mérida, Consejo Superior de Investigaciones Cientificas.
Bavuso, I., Furlan, G., Intagliata, E.E. and Steding, J. (2023) Circular economy in Roman and early medieval Europe – methods of analysis for a future agenda. *Open Archaeology* 9(1), 20220301. https://doi.org/10.1515/opar-2022-0301
Bessac, J-C. (2021) Les calcaires de Palmyre face aux autres roches de décoration architecturale et de sculpture. In R. Raja and J. Steding (eds) *Production Economy in Greater Roman Syria: Trade Networks and Production Processes,* Studies in Palmyrene Archaeology and History 2, 49–70. Turnhout, Brepols.
Cantineau, J. (1930) Textes funéraires palmyréniens, *Revue Biblique* 39.4, 520–551.
Cantineau, J. (1933) *Inventaire des inscriptions de Palmyre, fasc. IX: Le sanctuaire de Bêl.* Beirut, Imprimerie Catholique.
Collart, P. and Vicari, J. (1969) *Le Sanctuaire de Baalshamîn à Palmyre: Topographie et architecture, Le Sanctuaire de Baalshamîn à Palmyre I-II,* Bibliotheca Helvetica Romana X. Basel, Institut Suisse de Rome.
Cooley, A. (2000) The life-cycle of inscriptions. *Bulletin of the Institute of Classical Studies Supplement* 75, 1–5.
Dumser, E.A. (2018) Visual literacy and reuse in the architecture of late imperial Rome. In D. Ng and M. Swetnam-Burland (eds) *Reuse and Renovation in Roman Material Culture: Functions, Aesthetics, Interpretations,* 140–159. Cambridge, Cambridge University Press.
Gawlikowski, M. (1970) *Monuments funéraire de Palmyre,* Travaux du Centre d'Archéologie Méditerranéenne de l'Académie Polonaise des Sciences, 9. Warsaw, Editions Scientifiques de Pologne.
Gawlikowski, M. (2017) *Le sanctuaire d'Allat à Palmyre.* Warsaw, Wydawnictwa Uniwersytetu Warszawskiego.
Gawlikowski, M. (2019) The making of a city. In A.M. Nielsen and R. Raja (eds) *The Road to Palmyra,* 77–90. Copenhagen, Ny Carlsberg Glyptotek.
Hirt, A.M. (2021) Palmyra, Syria, and 'Imperial' Marble. In R. Raja and J. Steding (eds) *Production Economy in Greater Roman Syria: Trade Networks and Production Processes*, Studies in Palmyrene Archaeology and History 2, 101–122. Turnhout, Brepols.

Intagliata, E.E. (2017) Palmyra and its ramparts during the Tetrachy. In E. Rizos (ed.) *New Cities in Late Antiquity: Documents and Archaeology*, Bibliothèque de l'Antiquité Tardive 35, 71–83. Turnhout, Brepols.
Intagliata, E.E. (2018a) *Palmyra after Zenobia (AD 727-750): An Archaeological and Historical Reappraisal.* Oxford, Oxbow Books.
Intagliata, E.E. (2018b) The unpublished archival material from the Fonds d'Archives Paul Collart, University of Lausanne; remarks on the numismatic record of the Sanctuary of Baalshamin. *Rivista Italiana di Numismatica e Scienze Affini* 119, 15–30.
Intagliata, E.E. (2022) Reuse of building material and sculpture in late antique and early Islamic Palmyra (273–750 CE). An overview of the practice and several remarks on the evidence from the Sanctuary of Baalshamin. *Syria* 98, 419–438.
Intagliata, E.E. (2023) A matter of size. a dimensional approach to the study of reused inscriptions and sculptures from the sanctuary of Baalshamin. In N. Andrade and R. Raja (eds) *Exchange and Reuse in Roman Palmyra,* Studies in Palmyrene Archaeology and History 8, 99–112. Turnhout, Brepols.
Intagliata, E.E., Barker, S.J. and Courault, C. (eds) (2020) *City Walls in Late Antiquity: An empire-wide perspective.* Oxford, Oxbow Books.
Pallis, G. (2019) The second life of inscriptions in Late Antique and Byzantine Asia Minor: some remarks on the reuse of the inscribed material. *Gephyra* 18, 59–76.
Raja, R. (2022) *Pearl of the Desert: A History of Palmyra.* New York, Oxford University Press.
Raja, R., Bobou, O. and Romanowska, I. (2021) Three hundred years of Palmyrene history: unlocking archaeological data for studying past societal transformations. *PlOSONE* 16.11, e0256081. DOI: https://doi.org/10.1371/journal.pone.0256081.
Ruprechtsberger, E.M. (ed.) (1987) *Palmyra. Geschichte, Kunst und Kultur der Syrischen Oasenstadt*, Linzer Archäologische Forschungen Band 16. Linz, Gutenberg.
Russel, B. (2013) *The Economics of Roman Stone Trade.* Oxford, Oxford University Press.
Schmidt-Colinet, A. (2017) Die antiken Steinbrüche von Palmyra: Ein Vorbericht. *Mitteilungen der Deutschen Orient-Gesellschaft zu Berlin* 149, 159–196.
Schmidt-Colinet, A. and al-As'ad, W. (2013) *Palmyras Reichtum durch weltweiten Handel. Achäologische Untersuchungen im Bereich der hellenistischen Stadt.* Wien, Holzhausen Verlag.
Schnädelbach, K. (2010) *Topographia Palmyrena, I: Topography*. Damascus, DAI, Orient-Abteilung.
Seyrig, H., Amy, A. and Will, E. (1975) *Le temple de Bel à Palmyre*. Paris, Librairie orientaliste Paul Geuthner.
Sommer, M. (2018) *Palmyra: A History*, Cities of the Ancient World. Abingdon, Routledge.
Steding, J. (2023) Recarving of Palmyrene funerary portrairs. In N. Andrade and R. Raja (eds) *Exchange and Reuse in Roman Palmyra,* Studies in Palmyrene Archaeology and History 8, 59–74. Turnhout, Brepols.
Wielgosz, D. (2010) La sculpture en marbre à Palmyre. *Studia Palmyrénski* 11, 75–106.
Wielgosz, D. (2013) Coepimus et lapide pingere: Marble Decoration from the 'Baths of Diocletian' at Palmyra. *Studia Palmyrénskie* 12, 319–332.
Yon, J.-B. (2012) *Inscriptions grecques et latines de la Syrie 17.1. Palmyre*. Beirut, Institut Français du Proche Orient.

Appendix 1. Reused inscribed stones

No.	Language(s)	Object type	Original context	Dating original context	Reuse context	Dating reuse context	Main reference(s)
1	Palmyrene	column drum	Temple of Baalshamin	23	Temple of Baalshamin	Unknown	PAT 0167
2	Palmyrene	statue console	Temple of Baalshamin	32	Temple of Baalshamin	Unknown	PAT 0190
3	Palmyrene	column drum	Temple of Baalshamin	52	Temple of Baalshamin	Unknown	PAT 0168
4	Palmyrene	statue base	Temple of Baalshamin	60	Temple of Baalshamin	Unknown	PAT 0202
5	Palmyrene	statue console	Temple of Baalshamin	61	Temple of Baalshamin	Unknown	PAT 0192
6	Palmyrene	altar	Temple of Baalshamin	63	Temple of Baalshamin	Unknown	PAT 0179
7	Palmyrene	column drum	Temple of Baalshamin, Portico of Yarḥai	67	Temple of Baalshamin	Unknown	PAT 0159
8	Palmyrene	architrave	Temple of Baalshamin	67	Temple of Baalshamin	Unknown	PAT 0170
9	Palmyrene	column drum	Temple of Baalshamin	67	Temple of Baalshamin	Unknown	PAT 0171
10	Palmyrene	column drum	Temple of Baalshamin	70	Temple of Baalshamin	Unknown	PAT 0169
11	Palmyrene	architrave	Temple of Baalshamin	70	Temple of Baalshamin	Unknown	PAT 0187
12	Palmyrene	lintel	Temple of Baalshamin	70	Temple of Baalshamin	Unknown	PAT 0188
13	Palmyrene	column drum	Temple of Baalshamin, Portico of Alaisha	70	Temple of Baalshamin	Unknown	PAT 0161 F
14	Palmyrene	column drum	Temple of Baalshamin, Portico of Zabdilah	70	Temple of Baalshamin	Unknown	PAT 0162 A
15	Palmyrene	column drum	Temple of Baalshamin, Portico of Zabdilah	70	Temple of Baalshamin	Unknown	PAT 0162 B
16	Palmyrene	column drum	Temple of Baalshamin, Portico of Zabdilah	70	Temple of Baalshamin	Unknown	PAT 0162 C
17	Palmyrene	architrave	Temple of Baalshamin, Portico of Alaisha	70	Temple of Baalshamin	Unknown	PAT 0160
18	Palmyrene	column drum	Temple of Baalshamin, Portico of Wahbai	70	Temple of Baalshamin	Unknown	PAT 0163
19	Palmyrene	architrave	Temple of Baalshamin	70	Temple of Baalshamin	Unknown	PAT 0165
20	Palmyrene	stone block	Temple of Baalshamin	74	Temple of Baalshamin	Unknown	PAT 0186

(Continued)

No.	Language(s)	Object type	Original context	Dating original context	Reuse context	Dating reuse context	Main reference(s)
21	Palmyrene	stone block	Temple of Baalshamin	80	Temple of Baalshamin	Unknown	PAT 0185
22	Palmyrene	architrave	Temple of Baalshamin, Portico of Malkû	90	Temple of Baalshamin	Unknown	PAT 0164
23	Palmyrene	statue console	Temple of Baalshamin	90	Temple of Baalshamin	Unknown	PAT 0193
24	Greek/ Palmyrene	statue console	Temple of Baalshamin	98	Temple of Baalshamin	Unknown	PAT 0194; IGLS 148
25	Palmyrene	lintel	Temple of Baalshamin	100	Temple of Baalshamin	Unknown	PAT 0189
26	Palmyrene	stone block	Temple of Baalshamin	100	Temple of Baalshamin	Unknown	PAT 0235
27	Palmyrene	stone block	Temple of Baalshamin	100	Temple of Baalshamin	Byzantine	PAT 0230
28	Palmyrene	altar	Temple of Baalshamin	100	Temple of Baalshamin	Unknown	PAT 0181
29	Palmyrene	plinth	Temple of Baalshamin	100	Temple of Baalshamin	Unknown	PAT 0207
30	Palmyrene	stone block	Temple of Baalshamin	100	Temple of Baalshamin	Unknown	PAT 0218
31	Palmyrene	stone block	Temple of Baalshamin	100	Temple of Baalshamin	Unknown	PAT 0231
32	Palmyrene	column drum	Temple of Baalshamin	100	Temple of Baalshamin	Unknown	PAT 0173
33	Palmyrene	column drum	Temple of Baalshamin	100	Temple of Baalshamin	Unknown	PAT 0174
34	Palmyrene	statue console	Temple of Baalshamin	103/104	Temple of Baalshamin	Unknown	PAT 0196
35	Palmyrene	statue console	Temple of Baalshamin	103	Temple of Baalshamin	Byzantine	PAT 0195
36	Palmyrene	altar	Temple of Baalshamin	113	Temple of Baalshamin	Unknown	PAT 0183
37	Palmyrene	column drum	Temple of Baalshamin	120	Temple of Baalshamin	Unknown	PAT 0220
38	Palmyrene	column	Temple of Baalshamin	120	Temple of Baalshamin	Unknown	PAT 0222
39	Greek/ Palmyrene	statue console	Temple of Baalshamin	132	Temple of Baalshamin	Unknown	PAT 0197; IGLS 150
40	Palmyrene	stone block	Temple of Baalshamin	135	Temple of Baalshamin	Unknown	PAT 0223

(Continued)

No.	Language(s)	Object type	Original context	Dating original context	Reuse context	Dating reuse context	Main reference(s)
41	Palmyrene	statue console	Temple of Baalshamin	138	Temple of Baalshamin	Unknown	PAT 0198
42	Palmyrene	console	Temple of Baalshamin	150	Temple of Baalshamin	Unknown	PAT 0236
43	Palmyrene	statue console	Temple of Baalshamin	188	Temple of Baalshamin	Byzantine	Dunant 1971, 51
44	Palmyrene	altar	Temple of Baalshamin	207	Temple of Baalshamin	Unknown	PAT 0182
45	Palmyrene	statue console	Temple of Baalshamin	220	Temple of Baalshamin	Byzantine	PAT 0200
46	Palmyrene	statue console	Temple of Baalshamin	220	Temple of Baalshamin	Unknown	PAT 0201
47	Palmyrene	cornice	Temple of Baalshamin	273	Temple of Baalshamin	Unknown	Dunant 1971, 79
48	Palmyrene	cornice	Temple of Baalshamin	273	Temple of Baalshamin	Unknown	Dunant 1971, 91
49	Greek	cornice	Temple of Baalshamin	273	Temple of Baalshamin	Unknown	IGLS 147
50	Palmyrene	altar	Temple of Baalshamin	302	Temple of Baalshamin	Byzantine	Dunant 1971, 31
51	Palmyrene	stele	Funerary context	50	Temple of Baalshamin	Unknown	PAT 0210
52	Palmyrene	relief	Funerary context	200	Temple of Baalshamin	Unknown	PAT 0211
53	Palmyrene	relief	Funerary context	200	Temple of Baalshamin	Byzantine	PAT 0215
54	Palmyrene	stone block	Funerary context	273	Temple of Baalshamin	Unknown	Dunant 1971, 93
55	Palmyrene	altar	Temple of Allat	30	Temple of Allat	Unknown	Gawlikowski 2017, inscription 18
56	Palmyrene	architrave	Temple of Allat	55	Temple of Allat	Unknown	Gawlikowski 2017, inscription 4
57	Palmyrene	console	Temple of Allat	137	Temple of Allat	Unknown	Gawlikowski 2017, inscription 35
58	Palmyrene	door lintel	Temple of Allat	143	Temple of Allat	Unknown	PAT 1608
59	Palmyrene	stone block	Temple of Allat	273	Temple of Allat	Unknown	Gawlikowski 2017, inscription 7

(Continued)

No.	Language(s)	Object type	Original context	Dating original context	Reuse context	Dating reuse context	Main reference(s)
60	Palmyrene	stone block	Unknown religious context	62	Temple of Allat	Unknown	Gawlikowski 2017, inscription 30
61	Palmyrene	altar	Unknown religious context	273	Temple of Allat	Unknown	Gawlikowski 2017, inscription 22
62	Palmyrene	stone block	Unknown religious context	273	Temple of Allat	Unknown	Gawlikowski 2017, inscription 25
63	Palmyrene	stone slab	Funerary context	222	Temple of Allat	Unknown	PAT 1657
64	Palmyrene	console	Unknown	29	Temple of Allat	Unknown	Gawlikowski 2017, inscription 32
65	Palmyrene	stone block	Unknown	31	Temple of Allat	Unknown	Gawlikowski 2017, inscription 11
66	Palmyrene	console	Unknown	144	Temple of Allat	Unknown	Gawlikowski 2017, inscription 36; IGLS 127
67	Palmyrene	stone block	Unknown	273	Temple of Allat	Unknown	Gawlikowski 2017, inscription 29
68	Palmyrene	relief	Unknown	273	Temple of Allat	Unknown	Gawlikowski 2017, inscription 13
69	Palmyrene	unknown fragment	Unknown	Hellenistic	Temple of Bêl	32	PAT 1514
70	Palmyrene	stone block	Unknown	Hellenistic	Temple of Bêl	32	PAT 1515
71	Palmyrene	stone block	Unknown	Hellenistic	Temple of Bêl	32	PAT 1516
72	Palmyrene	statue base	Unknown	Hellenistic	Temple of Bêl	32	PAT 1517
73	Palmyrene	unknown fragment	Unknown	Hellenistic	Temple of Bêl	32	PAT 1518
74	Palmyrene	stone block	Unknown	Hellenistic	Temple of Bêl	32	PAT 1519
75	Palmyrene	column	Unknown	Hellenistic	Temple of Bêl	32	PAT 1520

(Continued)

No.	*Language(s)*	*Object type*	*Original context*	*Dating original context*	*Reuse context*	*Dating reuse context*	*Main reference(s)*
76	*Palmyrene*	*stone block*	*Unknown*	*Hellenistic*	*Temple of Bêl*	*32*	*PAT 1522*
77	*Palmyrene*	*stone block*	*Unknown*	*Hellenistic*	*Temple of Bêl*	*32*	*PAT 1523*
78	*Palmyrene*	*unknown fragment*	*Religious context*	*Hellenistic*	*Temple of Bêl*	*32*	*PAT 1521*
79	*Palmyrene*	*stone block*	*Unknown*	*Hellenistic*	*Temple of Bêl*	*32*	*PAT 1524*
80	*Palmyrene*	*statue base*	*Temple of Bêl*	*45*	*Temple of Bêl*	*Unknown*	*PAT 1347*
81	*Palmyrene*	*stone block*	*Temple of Bêl*	*60*	*Temple of Bêl*	*Unknown*	*PAT 1564*
82	*Greek/ Palmyrene*	*statue console*	*Temple of Bêl*	*120*	*Temple of Bêl*	*7th century*	*PAT 0267; IGLS 20*
83	*Greek/ Palmyrene*	*statue console*	*Temple of Bêl*	*127*	*Temple of Bêl*	*7th century*	*PAT 0266; IGLS 28*
84	*Palmyrene*	*stone block*	*Temple of Bêl*	*238*	*Temple of Bêl*	*7th century*	*PAT 1359*
85	*Palmyrene*	*altar*	*Religious context*	*273*	*Temple of Bêl*	*Unknown*	*PAT 2780*
86	*Palmyrene*	*unknown fragment*	*Funerary context*	*50*	*Temple of Bêl*	*Unknown*	*PAT 2774*
87	*Palmyrene*	*unknown fragment*	*Funerary context*	*273*	*Temple of Bêl*	*Unknown*	*PAT 0521*
88	*Palmyrene*	*statue base*	*Public Context*	*50*	*Temple of Bêl*	*Unknown*	*PAT 2773*
89	*Greek*	*stone block*	*Unknown*	*25*	*Temple of Bêl*	*Unknown*	*IGLS 1*
90	*Greek*	*stone block*	*Unknown*	*25*	*Temple of Bêl*	*Unknown*	*IGLS 2*
91	*Palmyrene*	*unknown fragment*	*Unknown*	*50*	*Temple of Bêl*	*Unknown*	*PAT 2775*
92	*Greek*	*statue base*	*Unknown*	*138*	*Temple of Bêl*	*Unknown*	*IGLS 4*
93	*Palmyrene*	*statue*	*Temple des Enseignes*	*62*	*Temple des Enseignes*	*Unknown*	*PAT 1941*
94	*Palmyrene*	*altar*	*Temple des Enseignes*	*273*	*Temple des Enseignes*	*Unknown*	*PAT 1152*
95	*Palmyrene*	*altar*	*Temple of Allat*	*115*	*Temple des Enseignes*	*Unknown*	*PAT 1929*

(Continued)

No.	Language(s)	Object type	Original context	Dating original context	Reuse context	Dating reuse context	Main reference(s)
96	Greek	stele	Funerary context	163	Temple des Enseignes	Unknown	IGLS 131
97	Palmyrene	lintel	Unknown	75	Temple des Enseignes	Unknown	PAT 1939
98	Palmyrene	unknown fragment	Unknown	143	Temple des Enseignes	Unknown	PAT 1938
99	Palmyrene	unknown fragment	Unknown	273	Temple des Enseignes	Unknown	PAT 2002
100	Greek	architrave	Unknown	273	Temple des Enseignes	Unknown	IGLS 142
101	Palmyrene	tablet	Funerary context	-12	Camp of Diocletian	293	PAT 1946
102	Palmyrene	tablet	Funerary context	18	Camp of Diocletian	293	PAT 1947
103	Greek/ Palmyrene	door lintel	Funerary context	115	Camp of Diocletian	293	PAT 0513
104	Greek/ Palmyrene	plaque	Funerary context	119	Camp of Diocletian	293	PAT 0515; IGLS 540
105	Palmyrene	plaque	Funerary context	139	Camp of Diocletian	293	PAT 0518
106	Palmyrene	door lintel	Funerary context	240	Camp of Diocletian	293	PAT 1945
107	Greek/ Palmyrene	unknown fragment	Funerary context, West necropolis	241	Camp of Diocletian	293	PAT 2823; IGLS 446
108	Greek	unknown fragment	Funerary context, West necropolis	273	Camp of Diocletian	293	IGLS 441
109	Latin	stone block	Military context	207	Camp of Diocletian	293	IGLS 118
110	Latin	stone block	Military context	217	Camp of Diocletian	293	IGLS 119
111	Palmyrene	statue console	Camp of Diocletian	273	Camp of Diocletian	293	PAT 1943
112	Latin/ Greek	stone block	Religious context	200	Camp of Diocletian	293	IGLS 133
113	Palmyrene	altar	Religious context	273	Camp of Diocletian	293	PAT 1936
114	Palmyrene	relief	Unknown	160	Camp of Diocletian	293	PAT 1933
115	Palmyrene	door lintel	Unknown	273	Camp of Diocletian	293	PAT 1644

(Continued)

No.	Language(s)	Object type	Original context	Dating original context	Reuse context	Dating reuse context	Main reference(s)
116	Palmyrene	cornice	Unknown	273	Camp of Diocletian	293	PAT 1651
117	Palmyrene	console	Unknown	273	Camp of Diocletian	293	PAT 1934
118	Palmyrene	lintel	Unknown	273	Camp of Diocletian	293	PAT 1940
119	Palmyrene	statue console	Unknown	273	Camp of Diocletian	293	PAT 2008
120	Palmyrene	stone block	Funerary context	86	Northern Wall	4th century	PAT 2816
121	Greek/ Palmyrene	lintel	Funerary context, North necropolis	155	Northern Wall	4th century	IGLS 467
122	Palmyrene	plaque	Funerary context	210	Northern Wall	4th century	al-As'ad, Gawlikowski and Yon 2012, no. 4
123	Palmyrene	plaque	Funerary context	241	Northern Wall	4th century	al-As'ad, Gawlikowski and Yon 2012, no. 2
124	Palmyrene	plaque	Funerary context	243	Northern Wall	4th century	al-As'ad, Gawlikowski and Yon 2012, no. 1
125	Palmyrene	stone block	Funerary context	273	Northern Wall	4th century	PAT 2822
126	Palmyrene	plaque	Funerary context	273	Northern Wall	4th century	al-As'ad, Gawlikowski and Yon 2012, no. 3
127	Palmyrene	stone block	Funerary context	273	Northern Wall	4th century	PAT 2818
128	Palmyrene	altar	Religious context	163	Northern Wall	4th century	PAT 1006
129	Palmyrene	altar	Religious context	189	Northern Wall	4th century	PAT 1005
130	Palmyrene	altar	Religious context	273	Northern Wall	4th century	PAT 1003
131	Palmyrene	altar	Religious context	273	Northern Wall	4th century	PAT 1004
132	Palmyrene	door lintel	Religious context	50	Justinian Wall	mid 5th century	PAT 0991

(Continued)

No.	Language(s)	Object type	Original context	Dating original context	Reuse context	Dating reuse context	Main reference(s)
133	Greek/ Palmyrene	door lintel	Funerary context, West necropolis	178	Justinian Wall	mid 5th century	PAT 1141; IGLS 436
134	Greek/ Palmyrene	lintel	Funerary context (Temple Tomb no. 175), North necropolis	212	Justinian Wall	mid 5th century	PAT 0557; IGLS 471
135	Greek/ Palmyrene	lintel	Funerary context, North necropolis	212	Justinian Wall	mid 5th century	PAT 1216; IGLS 473
136	Palmyrene	door lintel	Funerary context	273	Justinian Wall	mid 5th century	PAT 1586
137	Latin/ Greek/ Palmyrene	plaque	Funerary context	52	Rampart	4th century	PAT 2801; IGLS 535
138	Greek/ Palmyrene	plaque	Funerary context, North necropolis	99	Rampart	4th century	PAT 2817; IGLS 463
139	Palmyrene	relief	Temple of Nabu	100	Western Rampart	4th century	PAT 0008
140	Greek	plaque	Funerary context, North necropolis	144	Rampart	4th century	IGLS 466
141	Greek/ Palmyrene	plaque	Funerary context, North necropolis	164	Rampart	4th century	PAT 2819; IGLS 468
142	Greek/ Palmyrene	lintel	Funerary context, North necropolis	224	Military context	4th century	IGLS 477
143	Palmyrene	relief	Religious context	273	Southern Wall	4th century	PAT 1014
144	Greek/ Palmyrene	door lintel	Funerary context, West necropolis	67	Funerary context	4th century	PAT 1134; IGLS 402
145	Greek/ Palmyrene	door lintel	Funerary context, West necropolis	191	Funerary context	4th century	PAT 1135; IGLS 403
146	Palmyrene	altar	Religious context	217	Funerary context	Unknown	PAT 1445
147	Palmyrene	building block	Agora (Banquet hall)	30	Agora	Unknown	PAT 1429
148	Greek/ Palmyrene	pillar	Agora	164	Agora	Unknown	PAT 1372; IGLS 262
149	Palmyrene	stele	Funerary context	273	Byzantine cemetery	Byzantine	PAT 1897
150	Greek	column drum	Temple of Allat	100	Byzantine structure	Byzantine	IGLS 126
151	Greek	door lintel	Unknown	273	Byzantine church	Byzantine	IGLS 164
152	Greek	console	Unknown	182	Byzantine fortifications	Byzantine	IGLS 312

(Continued)

No.	Language(s)	Object type	Original context	Dating original context	Reuse context	Dating reuse context	Main reference(s)
153	Greek/ Palmyrene	column	Unknown	198	Byzantine fortifications	Byzantine	PAT 1063
154	Greek	stone block	Unknown	200	Byzantine fortifications	Byzantine	IGLS 308
155	Greek	statue base	Unknown	273	Byzantine fortifications	Byzantine	IGLS 314
156	Greek	altar	Temple of Baalshamin, Portico of Yarḥai	302	Byzantine structure	Byzantine	IGLS 154
157	Palmyrene	column	Temple of Bêl	128	Arab structure	7th century	PAT 1361
158	Palmyrene	altar	Religious context	138	Arab fortification	7th century	PAT 1432
159	Palmyrene	altar	Religious context	139	Arab fortification	7th century	PAT 1433
160	Palmyrene	altar	Religious context	139	Arab fortification	7th century	PAT 1472
161	Greek	statue console	Temple of Bêl	150	Arab structure	7th century	IGLS 10
162	Palmyrene	altar	Religious context	213	Arab fortification	7th century	PAT 1440
163	Palmyrene	altar	Religious context	219	Arab fortification	7th century	PAT 1442
164	Palmyrene	altar	Religious context	273	Arab fortification	7th century	PAT 1469
165	Palmyrene	altar	Religious context	273	Arab fortification	7th century	PAT 1470
166	Palmyrene	altar	Religious context	273	Arab fortification	7th century	PAT 1471
167	Palmyrene	altar	Religious context	273	Arab fortification	7th century	PAT 1473
168	Palmyrene	column	Temple of Bêl	273	Arab structure	7th century	PAT 1362
169	Greek	statue console	Temple of Bêl	150	Arab structure	7th century	IGLS 8
170	Palmyrene	door lintel	Funerary context	273	Gardens	Unknown	IGLS 2760
171	Greek	architrave	Unknown	273	Medieval Mosque	7th century	IGLS 298
172	Palmyrene	crater	Religious context	29	Unknown structure	Unknown	PAT 1553

(Continued)

No.	Language(s)	Object type	Original context	Dating original context	Reuse context	Dating reuse context	Main reference(s)
173	Palmyrene	architrave	Temple of Baalshamin, Portico of Yarḥai	67	Unknown structure	Unknown	Cantineau 1930, I,5/ Dunant 1971, 1B
174	Palmyrene	column	Religious context	67	Unknown structure	Unknown	PAT 1130
175	Greek/ Palmyrene	lintel	Funerary context	129	Unknown structure	Unknown	IGLS 541
176	Greek/ Palmyrene	statue console	Temple of Bêl	169	Unknown structure	Unknown	PAT 2748; IGLS 30
177	Greek	wall console	Agora	170	Unknown structure	Unknown	IGLS 204
178	Greek	statue console	Temple of Nabu	175	Unknown structure	Unknown	IGLS 184
179	Greek/ Palmyrene	door lintel	Religious context	203	Unknown structure	Unknown	PAT 0316
180	Palmyrene	door lintel	Funerary context	232	Source of Sérail	Unknown	IGLS 1788
181	Greek	console	Unknown	258	Unknown structure	Unknown	IGLS 55
182	Palmyrene	relief	Temple of Bêl	270	Unknown structure	Unknown	PAT 0063/0453
183	Palmyrene	console	Unknown	273	Unknown Late Antique structure	Unknown	PAT 1552
184	Palmyrene	column drum	Unknown	273	Unknown structure	Unknown	PAT 1583
185	Greek/ Palmyrene	lintel	Funerary context	251	Unknown structure	12th century	PAT 0558; IGLS 545

Plate 1 Textile fragment of a wall-hanging depicting a seated prince, 'Abbasid Egypt, 9th–10th century. Made from linen and wool; eccentric weft; tapestry weave. (The Textile Museum Collection, Washington, D.C., 721.14, Acquired by George Hewitt Myers in 1938. Photography by Breton Littlehales).

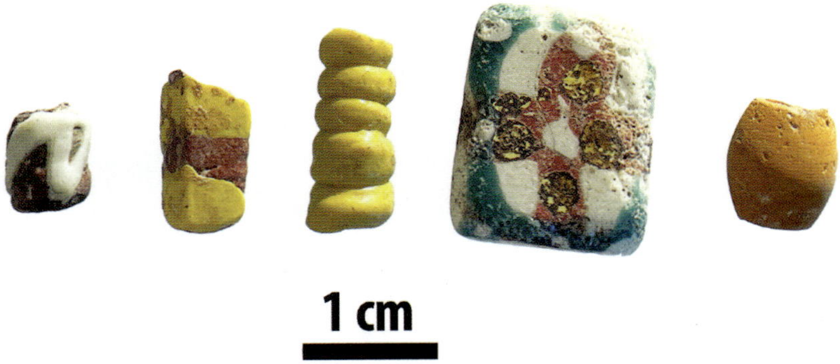

Plate 2 Beads from the Saint-Hilaire site, Huy, Belgium (Excavation AWaP, resp. S. de Bernardy de Sigoyer. Picture L. Van Wersch).

Plate 3 Pot n°. 1750, Vrijthof Cemetery, Maastricht, The Netherlands (Excavation Gemeente Maastricht. Picture L. Van Wersch).

Plate 4 Cobalt blue window glass, Stavelot, Belgium (Excavation AWaP, resp. B. Neuray. Picture L. Van Wersch).

Plate 5 Detail of the mosaic of the auricular apse, church of Germigny-des-Prés, France (Picture L. Van Wersch).

Plate 6 One of the niches of the nymphaeum of Segni (Rome). The mosaic decoration combines seashells, fragments of tufa stone, Egyptian blue pellets, stone chips and fragments of glass; end of the 2nd century BC (Photo by Cristina Boschetti).

Plate 7 Fragment of a carinated sealing-wax red glass cup and wall of an amber glass vessel reused in the mosaic-decoration of the House of the Bronze Bullock, Pompeii; late 1st century BC/ early 1st century AD (Photo by Cristina Boschetti).

Plate 8 Detail of the wall-mosaic decoration of the fountain of the Grand Duke of Tuscany, Pompeii. Five tesserae, forming a cross pattern, were obtained by cutting a fragment of an opaque sealing-wax red glass cup; AD 25–50 (Photo by Cristina Boschetti).

Plate 9 Fragment of the bottom of an opaque red glass plate, reworked to obtain a rectangular plaque; Denderah glass workshop, Egypt; middle of the 1st century BC. The Egyptian Museum, Cairo (Photo by Cristina Boschetti).

Plate 10 One example of the so-called Kimberley points, spearheads made by Aboriginals chipping glass bottles, in the Kimberley region, western Australia; 19th century. The Australian Museum, Sydney (Photo: @The Australian Museum).

Plate 11 Capped-mosaic glass beads and spindle whorls excavated in Egypt, Italy, Britain and Sweden, attributed to Egyptian workshops operating during the 5th century AD. Bead assemblage including (a) capped-beads from graves excavated in Qau (photo: @The Petrie Museum) and (b) drawing of a similar bead from the same graveyard (Brunton 1930, XLV); (c) roundish and (d) cylindrical capped beads excavated, respectively, from the cemetery of Campo Marchione, Italy (Boschetti et al. *2020) and from a grave in Norfolk (Heaser 2022); (e) spindle-whorl/sword bead from a house of Sandby borg, Sweden (Alfsdotter* et al. *2018); (f) spindle whorl from Antinoupolis, Egypt (Silvano 2017).*

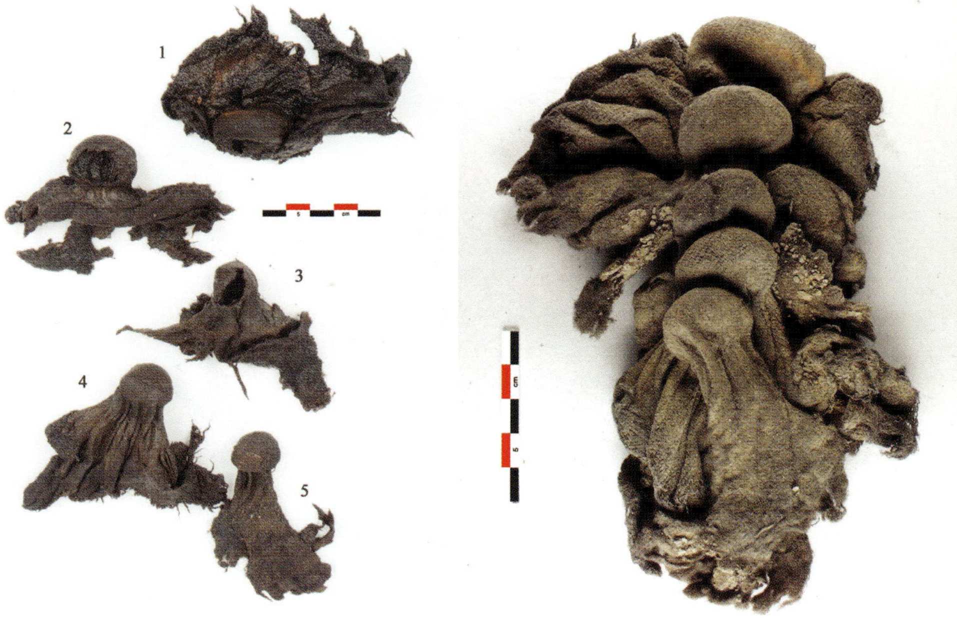

Plate 12 Tarring tools from Arles-Rhône 3 (Image after Médard et al. *2018, 400 fig. 3, reproduced with permission).*

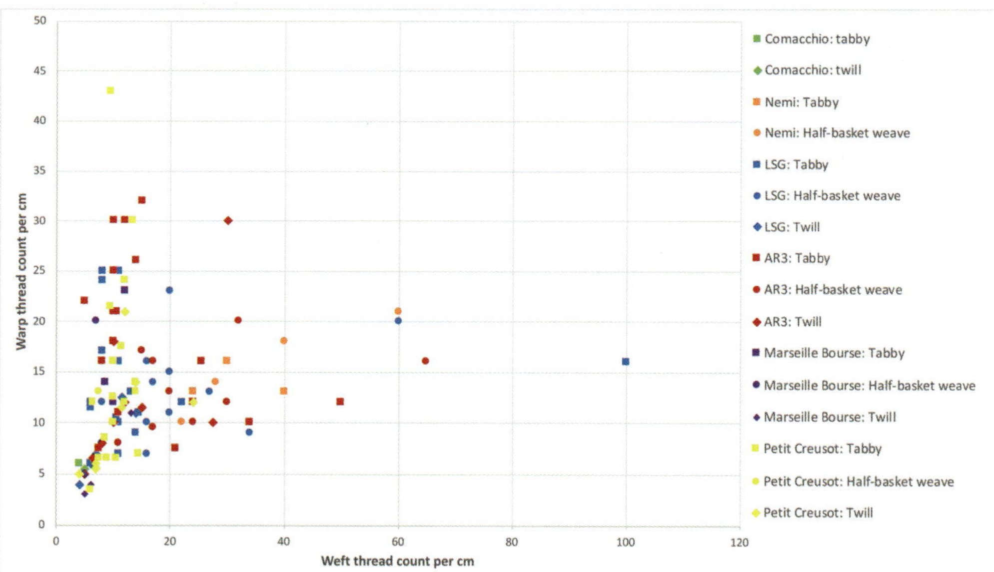

Plate 13 Scatter plot of warp and weft thread counts (per cm) of the textiles recovered from Comacchio, Nemi, Lyon-Saint-Georges 4, Arles-Rhône 3, Marseille-Bourse and Petit Creusot.

Plate 14 The Isle of Ely pendant, with rock crystal centre and garnet cloisonné details. The main body of the pendant measures approx. 3.3 cm in diameter (Image reproduced with the permission of The Museum of Archaeology and Anthropology, Cambridge. MAA 1951.565).

Plate 15 The gold and garnet pendant from 'the princess grave' (grave 42) at Street House (Copyright Stephen J. Sherlock. Image reproduced from Sherlock 2012, fig. 4).

Plate 16 The gold mounted Iron Age bead pendant from grave 43 at Street House. The bead mounted is of the Colchester 6a type. Image by David Currie, reproduced from A Royal Anglo-Saxon Cemetery at Street House, Loftus, North-East Yorkshire *by Stephen J Sherlock. (Sherlock 2012, plate 3.4).*

Plate 17 Star-shaped mould fragment ×1481.1. Photo by Sydvestjyske Museer (http://sol.sydvestjyskemuseer.dk/?mode=detail&genstandsnr=200331737&side=1&antal=18&indexno=1&search=200331737&sid=0e2632cb1415327771bfb54f4cbbc80f&tt=1) CC BY-NC-SA 2.0.

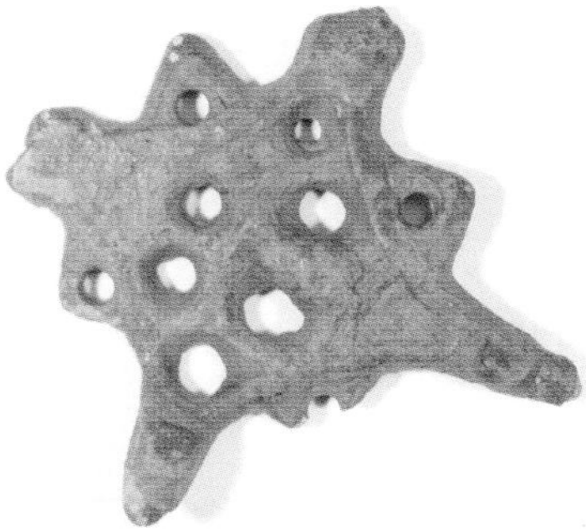

Plate 18 Star-shaped brooch example from Hviding, southwest Jutland. Item ID 200196713. Photo by Sydvestjyske Museer (http://sol.sydvestjyskemuseer.dk/?mode=detail&genstandsnr=200196713&side=1&antal=18&indexno=1&search=200196713&sid=0e2632cb1415327771bfb54f4cbbc80f&tt=1) CC BY-NC-SA 2.0.

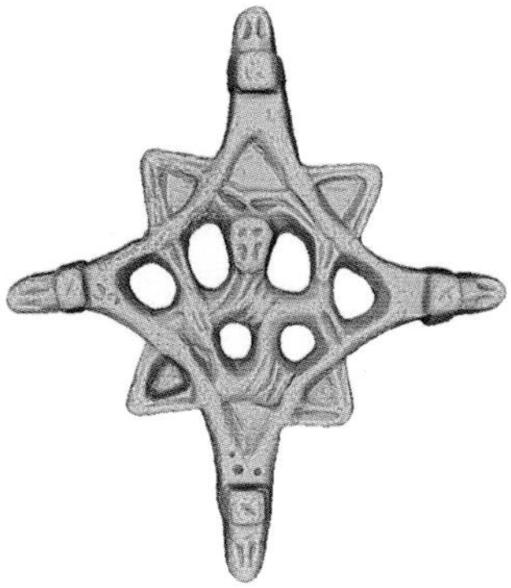

Plate 19 3D scan of star-shaped brooch example from Viby, Funen. Scan produced by kind permission of the National Museum of Denmark (Object ID number C41222).

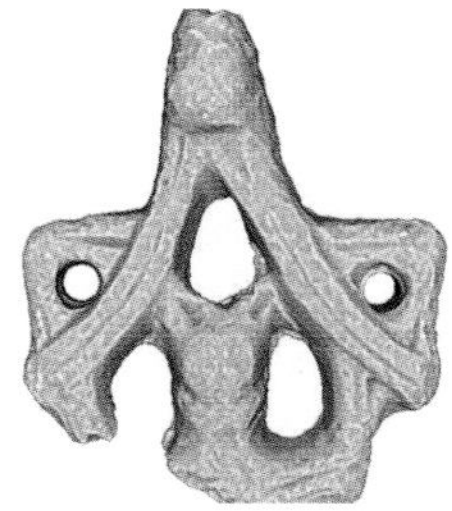

Plate 20 3D scan of star-shaped brooch example from Grenaa, eastern Jutland. Scan produced by kind permission of the National Museum of Denmark and DIME (Object ID number DI198107).

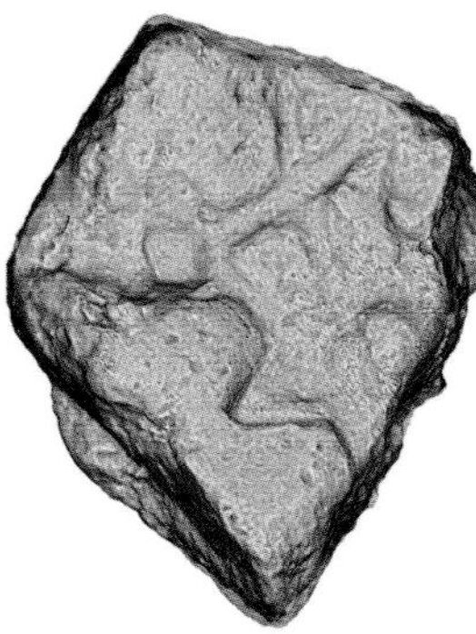

Plate 21 3D scan of mould fragment ×432.03. Scan produced by kind permission of Sydvestjyske Museer.

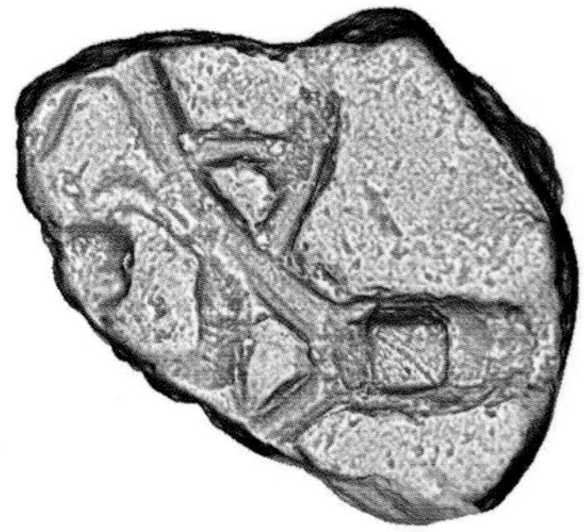

Plate 22 3D scan of mould fragment ×1481.1. Scan produced by kind permission of Sydvestjyske Museer.